Tears of the Cheetah

Tears of the Cheetah

And Other Tales from the Genetic Frontier

Dr. Stephen J. O'Brien

Foreword by Ernst Mayr

Thomas Dunne Books St. Martin's Griffin ⋒ New York

THOMAS DUNNE BOOKS.
An imprint of St. Martin's Press.

www.stmartins.com

Library of Congress Cataloging-in Publication Data

O'Brien, Stephen J.
    Tears of the cheetah : the genetic secrets of our animal ancestors / Stephen J. O'Brien ; foreword by Ernst Mayr.
        p. cm.
    ISBN 0-312-27286-3 (hc)
    ISBN 0-312-33900-3 (pbk)
    EAN 978-0312-33900-5
    Includes bibliographical references (p. 267) and index (p. 275).
    1. Animal genetics—Popular works. 2. Endangered species—Popular works. I. Title.

QH432.O276 2003
591.3'5—dc21   2003053164

First St. Martin's Griffin Edition: April 2005

10 9 8 7 6 5 4 3 2 1

To my teachers Ross MacIntyre, Bruce Wallace, and James Edwards, who introduced me to the wonder of genetic thinking, experimentation, and interpretation.

Contents

Foreword

THERE IS NO BETTER WAY TO INTRODUCE A BEGINNER into the achievements of molecular biology than Stephen J. O'Brien's *Tears of the Cheetah*. Some early biochemists loudly proclaimed that molecular biology would exterminate the rest of biology. "There is only one biology," one of them said, and "This is molecular biology." Nothing could have been more wrong. Instead, the study of molecules has enormously enriched organismic biology and led to astonishing discoveries in just about all branches of biology. O'Brien, in fourteen chapters, shows in case study after case study how the molecular discoveries of genomes have shed unexpected light on such problems as the loss of genetic variation in the cheetah, the Florida panther, and the Asian lion; how we can find the degree of genetic difference among populations of whales, of Indonesian orangutans, and of many species of uncertain taxonomic rank; why the unique mating system of lions is not in conflict with kin selection; how AIDS originated, and why it is so resistant to all medical effort. Immune genes tell us about historical epidemics hundreds or thousands of years ago. These fascinating scenarios are told by O'Brien with such literary mastery that one can hardly lay his book down.

In chapter 10 the extraordinary similarities of the genomes of different mammals (including the human) are presented, and it is pointed out to what great extent the solution of human medical problems is helped by comparative studies of the genomes of other mammalian families—for instance, cats. Though often dealing with highly technical matters, O'Brien has succeeded in presenting his stories in a simple language that can be understood even by the nonexpert. And he also persuades us how important the findings made on animals often are for human medicine. There is no other book I have read in recent years from which I have learned more, and which I have more enjoyed, than this one. And so will every reader.

—Ernst Mayr,
Alexander Agassiz Professor of Zoology,
Emeritus, Harvard University

Prologue

THE BOOK YOU HAVE OPENED CONTAINS A COLLECTION OF stories. They are adventures, they are science mysteries, they are medical enigmas, and they are detective stories. Most draw their theme around an inimitable threatened animal species and scientific advances that unveil past and present dangers. The chronicles are all true and in different ways illustrate the power of new genomic technologies to uncover hidden secrets in the history of wildlife species, of companion animals, and of ourselves.

At first glimpse these tales describe the peril of beloved endangered species—cheetahs, humpback whales, giant pandas, and others. What we discovered beneath the surface of the fragile wildlife species reveals the rationale and context for their successes, for their survival, and for their vulnerability. The new perspectives derive from reading genetic codes of living species, only recently available for inspection. The spectacular developments in genetic technology and gene identification can be applied to any species with surprising and sometimes disturbing results. From the viewpoint of evolutionary hindsight, a thousand fables are slowly emerging through the lens of modern genomic enquiries. What began as a quest for reversing species extinction opened my eyes to as rich a history of magnificent creatures as anyone could have imagined. The genetic thread that strings these stories together is the indisputable unity of all living things. We are all intricately joined by a vast, weblike genealogical network to our most ancient ancestors.

The genomes of modern mammals, the sum total of an individual's genetic instructions, contain an extraordinary cache of information. In their genetic endowments of three billion nucleotide letters reside thirty-five thousand to fifty thousand genes that specify each creature's development. But they also retain the script of historic brushes with extinction, adaptation, and survival. Every day nature's dutiful field experiments test new gene variants across individuals in populations, among species, and across geographical space. All these trials have been neatly catalogued in the genetic sequence of the survivors, the plants and animals alive today. Scientists are only beginning to interpret the genetic "pawprints" of ancient events. We are learning as we go how to discern the message and lessons in DNA. Though progress may seem slow, the early views on this genetic safari have galvanized our thinking and optimism for solving countless mysteries about how living things came to be.

The genomics era appears to have unprecedented promise and potential for nearly every aspect of biology. It is as if the printing press were just invented and we anticipate the wide dissemination of the thoughts, ideas, feelings, and experiences of a whole generation. The difference is that our collective genomes encode the lessons of not one but tens of thousands of writers, multiplied across every evolutionary lineage while also possessing their own unique and defining experience. As the periodic table empowered chemical design and the silicon chip changed all things computational forever, so will our cracking the genome be remembered as a turning point in the underpinning of biological happenings—past, present, and future.

An underlying theme of the science mysteries is the unusual insight that animal studies bring to human medicine. Wild species have no hospital emergency rooms, no HMOs or pharmacies to treat their ills. Still they are regularly assaulted with scourges nearly identical to those that afflict humans—cancers, deadly infectious diseases like AIDS and hepatitis, degenerative diseases such as multiple sclerosis, Alzheimer's, and arthritis. Many victims succumb and die; indeed, well over 99.9% of all mammalian species that have walked the earth have gone extinct. But others have escaped, and those lucky species survive quietly today, carrying the evolved secrets to their success in their genetic endowment. Can medical science learn from natural solutions to hereditary, infectious, and neoplastic diseases acquired by free-living orangutans, lions, cougars, and barn mice? I believe we can, and I will illustrate how by the examples in the coming chapters.

These stories offer a window into twenty-first-century science, where undoubtedly countless biomedical advances will come from mining the genome. These are parables of hope and lessons of survival. They also navigate through the torturous but exhilarating process of scientific discovery, interpretation, and policy development.

Each chapter tells its own special narrative with twists and turns no novelist would have imagined. I selected them because I have been personally involved with each and with the lead characters I describe. The cast of players are scientists, graduate students, postdoctoral fellows, physicians, veterinarians, field ecologists, and many others who play into the mix of research advancement. I have had the privilege of joining these adventures as a government scientist, employed as a geneticist for thirty years at the U.S. National Institutes of Health (NIH).

In 1971 I arrived as a rather naïve *Drosophila* (fruit fly) geneticist, wondering if my chosen discipline might provide any benefit to medical research. I now serve as a laboratory chief in NIH's intramural research program, overseeing the research projects of students, fellows, and senior staff. We strive to use the genetic technologies and advances to chip away at the causes, diagnostics, and treatments for cancer and infectious disease. The stories I tell come from the challenge and exhilaration of exploring into the deepest mysteries of biology—why some species have survived while others have not.

In the end, we have been fortunate to uncover animal and human genomic secrets. Some discoveries, like solving the inscrutable origins of giant pandas, resolved dusty old academic conundrums. Others, like the fragility of the cheetah and Florida panther, informed the workings of conservation plans. Still others, like the AIDS plague in lions and in humankind, opened new avenues for clinical therapy for today's most devastating infectious disease killer.

This collection offers a brief peek at the brilliant landscape that the post-genomics era will display, a fascinating view of biological fits and starts that preview finely tuned advances in conservation, forensics, and medicine. My hope is to render the process understandable to the interested reader with little background in the jargon and mystery of genomic thinking. In the process, I have used some tricks, analogies, and literary license to make principles understandable. Throughout, the technical terms are kept to a minimum. The glossary at the back explains some of the basic terminology. I hope the reader will enjoy the learning and share in the wonder of amazing science that is opening our eyes for discovery and application across all the disciplines of modern life sciences.

Tears of the Cheetah

One

A Mouse That Roared

IT IS CONSIDERED CHINA'S GOLDEN AGE. FROM A.D. 960 TO 1279 the Song Dynasty ruled over a cultural and technological renaissance that gave rise to seminal discoveries—printing, magnetic force, the compass, and gunpowder—several hundred years prior to parallel innovations in Western Europe. The Song and succeeding Ming dynasties saw a dazzling population expansion that doubled the 80 million Chinese citizens in the fourteenth century to over 160 million by the peak of the Ming Dynasty in 1650. Today, 1.3 billion people live in China.

The population boom was marked by urban sprawl and agricultural improvements. Crops and grain supplies increased steadily through the Song and Ming dynasties with expansive cultivation, growth in productivity per acre, and a geometric increase in irrigation. Through the last millennium China blossomed as an agrarian culture with 90% of its land under cultivation and less than 2% dedicated to pastures for animal grazing.

The stores of grain across China provided a ripe opportunity for rodents, particularly mice and rats as they became what scientists call commensal species—animals that flourish alongside human activities. Dogs, cats, houseflies, mosquitoes, cockroaches, and pigeons are all commensal species. Wild mice do particularly well in barns, silos, and grain stashes, producing litters of up to a dozen pups each month. As China's agriculture flourished, so did its mice, which prospered to numbers in the millions, if not billions.

Then, some time deep in the Middle Ages, a ferocious disease outbreak devastated Chinese mouse populations. Such a plague would likely have been welcome to the peasants and farmers of the time. To the scientists who stumbled upon this obscure history centuries later, the plague would reveal itself as an unusual and profoundly significant evolutionary event. What they uncovered about these mice would change the way biologists and medical detectives perceived the history of wild species.

The devastating virus that hit these mice caused blood cancers, lower limb paralysis, and paraplegia. By the thousands these overcrowded, stressed mouse communities would succumb to the fatal disease. No one is certain where the virus came from—maybe from domestic cats brought in to dispatch the barn mice, or from birds or livestock. The epidemic killed tens of millions of mice before the tide changed. Somehow certain mice survived and continued to reproduce, feed, and spread, once again to be limited only by available food, predators, and opportunities to set up a nest. It was a narrow escape.

How did the survivors avoid the menacing virus that felled so many mice? The puzzle was solved centuries later by a seasoned medical pathologist whose curiosity and scientific acumen uncovered the eerie scenario as if peeling away the layers of an onion until the prize core, the puzzle's solution, was his.

Dr. Murray Gardner spoke softly but deliberately to the high school boys just after midnight. "The key is stealth. Nothing is illegal here, but be very careful not to be seen and talk to no one. The last thing we need is to wind up in the newspapers; that would stifle the whole operation."

The boys were amateur mouse catchers, equipped with cotton work gloves, dark plastic garbage bags for the take, and miner's helmets with stun lights to freeze/terrorize the mice so they could grab them and toss them in their sacks. The hunting grounds: a squab farm near Lake Casitas in south Ventura County, California, forty miles north of Los Angeles. Squabs are pigeons raised as a delicacy for Chinese restaurants, and the farm had ten thousand of them brooding in small pens. Under the dung-saturated pen beds lived hundreds of house mice, *Mus musculus domesticus*, quietly pilfering the squab's grain feed.

Gardner needed the mice to search for new retroviruses, a nasty type of virus that causes cancers, notably leukemias and lymphomas in chickens, cats, and mice. Retroviruses are unusual in that their genes consist of RNA—nucleic acids that control cellular activity—rather than standard DNA. These viruses use an enzyme to copy their RNA genetic code into DNA form, which it then inserts into its victim's DNA. RNA is usually a product of DNA, not the other way around—hence the prefix *retro*. Murray Gardner was keen to sample wild mice because like humans, but unlike most lab mice, they had not been bred artificially to shed or minimize genetic diversity. The mouse retroviruses isolated before Murray's hunts all came from inbred strains, mice descended from twenty or more generations of brother-sister incestuous matings.

Murray had the farmer's permission to collect the mice and take them to his laboratory, but only if he kept the operation quiet. The reason for the secrecy was simple: Squabs are raised for human consumption and farms that raise them face a monthly inspection by the State of California Rodent Control Board to certify that the farms are "rodent free." Once the mice were exposed, forced traps and poison would be inevitable. So the boys collected the mice quickly and before dawn dropped their catch in a drum at a rural gasoline station to be intercepted by Gardner.

Murray Gardner is a curious and sometimes impatient fellow. In another life, he may have been a Sherlock Holmes–style detective, a Greek philosopher, or even a charismatic statesman. A real-life Hawkeye Pierce, Murray served as a MASH doctor in the Korean War, patching injuries and administering obstetric care to frightened young Asian women who carried GIs' babies. After the war, he trained as a medical pathologist and joined the University of Southern California faculty as a medical researcher in 1964.

By 1970, Murray was forty-one, professionally successful, and well settled in his academic routine. His clinical and teaching responsibilities were important but not particularly challenging. In his spare time he consumed scientific articles on cancer, infectious diseases, and medical advances with demonic passion. His first research project was an attempt to show that Los Angeles smog caused cancer in lab mice. He had heard about President Nixon's War on Cancer, a kind of "moon shot" ambition to understand and cure the disease. The program infused millions of government dollars from the U.S. National Institutes of Health (NIH) into targeting causes, diagnostics, and new treatments for cancers. A large part of the new cash went to the Virus Cancer Program, an expansive effort launched by the National Cancer Institute (NCI) to discover human viruses that would cause cancer.

The Virus Cancer Program, which lasted from 1968 to 1980, was sadly short-lived because no human cancer-causing viruses were immediately discovered and critics succeeded in pulling the financial plug by arguing that viruses had nothing to do with human cancer. Today we know that several human viruses cause cancers responsible for hundreds of millions of deaths. Papilloma viruses are the prime cause of cervical cancers; hepatitis B virus leads to liver cancer, affecting 300 million people across the world; and HIV leads to lymphoma, Kaposi's sarcoma, and other tumors in AIDS patients. In retrospect, the Virus Cancer Program was hardly misguided but cutting edge—a research effort ahead of its time.

Murray Gardner was busy exposing laboratory mice to smog and auto emissions on interchanges of Los Angeles freeways when Robert Huebner, a leader of NCI's Virus Cancer Program, asked for his help. Huebner's research group had identified many cancer-causing retroviruses in lab mice, but he worried that the intense inbreeding of the lab mice had compromised their discovery.

Huebner knew that outbred species like humans and wild animals had some thirty-five thousand genes as their genetic base and that nearly every gene had some level of genetic variation. If such diversity includes genes that specify immune response to viruses or other infectious diseases, then the inbreeding of laboratory mice may have inadvertently eliminated key genetic regulators for virus replication and virulence. This hunch would offer a plausible explanation for the ease with which multiple tumor viruses were harvested from inbred mice and chickens, while none had been found so far in humans. Perhaps retroviruses were really present in humans, but were actively repressed by our genetic diversity. The idea made evolutionary sense because virus-repressing genes would provide a real benefit for outbred species: the prevention of viral-induced cancers.

Huebner and Murray reckoned that such cancer-causing viruses might be lurking in outbred species, but suppressed by the species' genes into a latent form. They agreed that a search for such agents in the wild might uncover some very interesting natural microbes in a few rare sensitive mice, ones that were the forebears of the tumor viruses that had been discovered in lab mice. All they needed was to get their hands on some wild mice and take a look using standard virological tools.

The hunt was on—at dairy farms, racetracks, aviaries, alleys, birdseed factories, freeway trellises, and any place wild mice might be lurking. Somewhere between ten thousand and twenty thousand mice were collected by Gardner's clandestine operations over the next decade. He paid the boys ten cents a mouse. After scores of false starts, narrow escapes, and raised eyebrows, Gardner managed to capture mice from fifteen locales in the greater L.A. area. He watched them age and searched for cancer, retroviruses, and other viral diseases. Nearly all the mice were free of cancer, retroviruses, or other infections except for a few sporadic tumors in aging mice. Nor did Gardner find the natural microbes they thought might echo tumor viruses in lab mice. The exception was the squab farm near Lake Casitas.

The mice nestled in the squab bins were different. They were battling a massive epidemic of lethal retroviruses. Nearly 85% of the farm mice carried evidence of exposure to one of two ravaging viruses. The more virulent strain was called ecotropic MuLV, murine (i.e., mouse) leukemia virus. *Ecotropic* means that the virus grew in mouse cells in the lab, but not in cultured cells from other species like human, rat, or cat. Murray isolated and injected the wild mouse virus into lab mice; it caused a fatal spinal hind limb paralysis and in older animals a blood cancer called "lymphoma." The spinal paralysis was termed spongiform polio-encephalomyelopathy, and it developed in the wild mice as early as ten months after birth. The virus killed the wild mice it infected and was transmitted to offspring through breast-feeding.

At last, a wild mice population laced with a fatal retrovirus had been found. Gardner could now study a cancer-causing virus in a population that more closely resembled the genetic diversity of humans and other outbred species.

In scientific inquiry, it seems that the more you discover, the more new unanswered questions appear. Murray sampled the Lake Casitas mouse population monthly for several years, fully expecting the paralytic virus to sweep through the pens and extirpate the mice. That did not happen. The mice flourished at high density for a decade and disease incidence stayed at around 15%. Infected mice died quickly, but 85% of the mice never succumbed to the virus and were never paralyzed. There seemed to be a powerful check on the impact of the deadly virus within the population.

I first heard this curious saga of the lucky mice from Murray at the 1977 Annual Conference of the Virus Cancer Program. The apparent freeze-frame nature of the epidemic was tantalizing. How could a generic mouse population tucked beneath a squab bed in rural California survive a lethal epidemic indefinitely? To me it seemed like a genetic difference, but in those days we were more used to thinking of variable genes as predispositions to hereditary genetic disease like sickle cell anemia or cystic fibrosis. A gene that could block a fatal virus infection would be particularly interesting, so together we set out to find it.

Murray was anxious to enlist my help because in the early 1970s there were few geneticists who also studied retroviruses. My training as a *Drosophila* (fruit fly) geneticist at Cornell exposed me to experiments in genetic transmission plus gave me a foundation in population and evolutionary genetics. Population genetic experts were tracking fruit fly populations in cages or in natural settings to view patterns of genetic variation that were the precursors to species adaptation and species formation. My postdoctoral fellowship at NCI brought me shoulder to shoulder with the giants of retrovirology, who showed me the fascinating complexity of tumor virology. Murray asked me to help him find out if Lake Casitas mice harbored a resistance gene to defend against the deadly retrovirus and, if so, to figure out how it worked.

To get at the explanation, we needed first to prove the resistant mice really had such a gene and could transmit it to their progeny. We were not sure how to do this, but we had some tricks from classical genetics that I learned from working with fruit flies. Murray was an attentive and enthusiastic student.

Early on, Murray discovered that Lake Casitas paralytic virus was genetically related to another laboratory retrovirus, one that caused a very high cancer incidence in an inbred mouse strain called AKR. This connection would soon prove to be the key that allowed us to get at the Lake Casitas wild mouse resistance.

AKR mice had been inbred extensively in the 1920s and selected by their breeders for a high incidence of leukemia, so much so that 100% of each generation develop leukemia and die before their first birthday. Some rather elegant experiments in molecular biology in the early 1970s showed why. AKR mice carried three full-length copies of a murine retrovirus, termed *AKV* (for A̲K̲R̲ v̲irus), in their chromosomes, nestled between the normal mouse genes for making a mouse. These three AKV genomes (a genome is a full-length copy of the viruses' genetic endowment) are passed down from parents to offspring on the chromosomes of sperm and eggs just like all the regular genes. In effect, the viruses are hitchhikers on the mouse's chromosomes.

After birth, something triggers these latent endogenous viruses to start replicating themselves and spread through lymphocytes (white blood cells). Endogenous means the virus lives in the host chromosomes and is passed vertically to offspring, in contrast to exogenous viruses like flu or smallpox, which spread horizontally between individuals. What these released endogenous retroviruses do best is cause leukemia. They accomplish this feat by infecting a lymphocyte and inserting themselves into a chromosome adjacent to one of a few hundred mouse genes that, when expressed in the wrong cell, lead to uncontrolled cell division, or cancer. The AKR virus simply activates the adjacent gene by providing the genetic equivalent of a "turn me on!" signal to the cellular machinery that decides which genes get expressed. The process transforms the infected lymphocyte into a wildly uncontrolled dividing cell, the first step in leukemia. The AKR mice have high virus titers (concentrations) in their blood and die of cancer as a consequence. Could the Lake Casitas (LC) resistant mice also block the AKR virus? If so, then we could conclude that the wild mice must have carried an anti-retrovirus resistance gene.

Murray set up two mating crosses, one between AKR mice and the LC mice that were infected with the wild mouse virus and a second cross between AKR mice and LC mice that were virus-free and presumably resistant to the virus and its paralysis. The first cross produced offspring that developed AKV virus production, leukemia, and early death, as in their AKR parents. But the second cross was very different. Several virus-free LC parents, when crossed to AKR mice, produced dozens of offspring with no viremia, no leukemia, no paralysis, and long life. Something carried in the chromosomes from the LC virus-free mice had completely shut down the AKV virus in the hybrid offspring. It had to be a resistance gene in the LC resistant mice that neutralized not only the wild mice virus, but also its close relative, the AKV nested in mouse chromosomes.

Other virus-free LC parents produced two categories of offspring in about equal proportions: one group of pups was highly viremic (that is, riddled with virus particles throughout their bodies and bloodstream) and the second completely virus-free. This pattern made sense to a geneticist. The virus-free LC mice with no viremic offspring carried two copies of the wild mouse resistance gene; the LC virus-free mice that produced both viremic and virus-free offspring had one resistance gene and one sensitive gene.

To nail our conclusion further, Murray then crossed the resistant AKR-LC hybrid progeny back to their AKR parents. This mating produced a 50:50 ratio of virus-free mice to viremia/leukemia-riddled offspring. That 50:50 split is precisely the prediction of Gregor Mendel's first law of genetics, the law of single gene segregation. The crosses left no doubt in our minds that the LC mice had a genetic gold nugget on one of their chromosomes, a powerful force that protected its carriers from retroviremia, leukemia, and early death. Murray and I named the new gene *AKVR* for "AKV restriction."

So how does this retrovirus restriction gene actually work? The answer came once the gene was mapped to a specific mouse chromosome position and then isolated using a process called gene cloning. When we looked at the DNA sequence of the isolated mouse gene, we were stunned at what it was. The so-called restriction gene turned out to be a miniature but foreshortened version of the retroviral genome that was causing the disease in LC and AKR mice.

Retroviral genome sequences are rather simple structures of around nine thousand nucleotides (nucleotides, or base pairs, are the DNA letters of the genetic code that string together the genes) that specify four genes: *env,* which encodes a surface envelope protein that binds the virus to the cells it infects; *pol,* which specifies a polymerase enzyme to make DNA copies of the virus's RNA genome; *gag,* a gene that makes an internal virus core protein to encompass its fragile RNA; and *LTR,* a velcro-gene that sticks the virus DNA copy to the host's DNA to facilitate its insertion within the huge animal host chromosome. The LC mouse restriction gene, *AKVR,* is a shortened version of the virus that makes one good gene product of the four, the envelope protein on the virus's outer surface.

Since the LC gene is an incomplete virus, it cannot cause leukemia like the complete endogenous retrovirus carried by the AKR mice. All the *AKVR* did was pump out retroviral envelope proteins in the white blood cells of mice. So how did it protect them from the deadly LC virus? The answer became clear once we understood the way retroviruses cause leukemia in the first place.

Retroviruses first enter cells by recognizing specific receptors or doorways on the cell surface of the tissues they infect. Different retroviruses require different receptors, and mouse viruses like the LC virus or AKV use a receptor called Rec-1, a large snakelike protein embedded within the cell's membrane. The receptor contains short protein buds extending out of the membrane like fingers. One of these fingers has a lock-and-key recognition signal that binds like a magnet to retroviral envelope proteins, causing the cell's membrane to dissolve and allowing the virus to inject its DNA into the cell.

ed cells soon become tiny factories producing new viral parts,
ting viral genes, and assembling new viruses. In the process,
some newly synthesized wayward envelope proteins make their way
to the cell surface, where they bind strongly to their complementary
receptors. These envelope proteins cover the receptors, rendering
them unavailable to a new infection by related viruses floating by in
the bloodstream. That virus-mediated prevention of secondary infection has a name, *viral interference,* and in a weird aberration it lies at
the crux of the genetic restriction we saw in the Lake Casitas mice.

The resistant mice keep the paralytic virus at bay because the requisite cell receptor, Rec-1, is saturated by viral envelope proteins produced not by a virus but by the *AKVR* restriction gene of the LC
resistant mice. It is an exquisite solution, a genetic vaccine against a
fatal virus. The mouse makes an innocuous virus envelope protein
that doesn't hurt them and quite effectively blocks a fatal LC or AKR
virus attack. But, we wondered, where did this magic gene come
from? How did it get there? And do other mice carry it?

The mice in the Lake Casitas farm that resisted virus infection and
paralysis all had the *AKVR* gene, and, not surprisingly, none of the
virus-infected paralytic LC mice had the gene. Other mouse populations around California and other states lacked both—no retrovirus
and no *AKVR* gene. The LC mice were alone in their battle with the
fatal virus and unique in their possession of the *AKVR* gene. Or so it
seemed.

The next twist in the story came from an unexpected place: virology
research in a subspecies of wild mouse in Japan—*Mus musculus
molossinus*. A subspecies is a population of animals that are genetically and visually distinct from other populations of the same
species, such as Siberian and Bengal tigers, or grizzly bears and
Japanese Hokkaido brown bears. A group of Japanese virologists had
discovered that their wild mice had a relatively high prevalence of
viremia just like the LC mice and an apparent genetic restriction situation as well. Murray and I sent for some of these Japanese mice
and bred them to Lake Casitas resistant mice. When the hybrid mice

were mated to AKR mice, all the offspring were virus-free, so the hybrids were restrictive for the AKV viremia and leukemia. A geneticist knows this means that exactly the same retroviral restriction gene was present in both the LC and the Japanese mice, a key discovery in the puzzle. Molecular analysis uncovered the *AKVR* gene, the same half-complete retroviral genome sequence, in the Japanese mice as well, leading us to suspect that the LC mice may have some sort of unbeknownst kinship or ancestry with Japanese wild mice.

Further sleuthing uncovered research studies on the migration patterns of wild mice in Asia. Scientists looking at the patterns of DNA variation in various mouse genes had shown that all the mice in Japan today were descendants from recent immigrants to the island. Northern continental Asia (Russia, Mongolia, and northern China) is populated by a mouse subspecies called *Mus musculus musculus*, while southern China and Southeast Asia has a different mouse subspecies, *Mus musculus castaneous*. It turns out that two rather recent migrations, one from the north in the sixteenth century and a second from the south in the eighteenth century, were the forebears of modern Japanese mice. *Mus musculus molossinus* is a hybrid subspecies between the northern *musculus* and the southern *castaneous*. We looked at north Asian *musculus* and found no virus and no *AKVR* restriction gene, but the southern *M. m. castaneous* had both. The *AKVR* restriction gene and the LC virus were quietly living in southern Asia in a situation identical to the squab farm near Lake Casitas.

By piecing together these related observations, we now can imagine the origins of the Lake Casitas restriction gene. Our story takes us back to the Song Dynasty epidemic that ravaged the Chinese mouse population. As must have happened countless times, an infected female mouse became pregnant and immediately transmitted the virus to its embryo. The virus attempted to integrate into the embryonic cell's chromosome, but this one time, it made a mistake, losing half of the virus genes and placing the other half into the mouse's twelfth chromosome. As the embryo developed into a pup, every cell descended from the infected cell carried the new retroviral *env* gene segments. Fortuitously for this mouse and its descendants, the foreshortened virus genome had integrated into a chromosome region that happened to contain a switch, a genetic regulatory element that can flip on virus genes in lymphoid blood tissues. That baby mouse was magically resistant to the raging epidemic into which it was born. S(he) grew up, mated, and transmitted the resistance gene to offspring who would carry the same blessing. The epidemic provided a strong natural selective influence, favoring carriers of the newly created *AKVR* gene. Mice with the *AKVR* gene lived and reproduced; those without it died. When Chinese mice migrated to Japan in the sixteenth century, the retrovirus and *AKVR* restriction gene, locked in a virus-host balancing act, went with them.

We now believe the Lake Casitas mice represent an American branch of the same phenomenon. Murray's California farm was first settled in the nineteenth century by Chinese immigrant farmers, whose descendants maintain it today. The immigrants carried squabs destined for California's Chinese cuisine across the Pacific, and as an unsuspected bonus they also brought mice, the paralytic retrovirus, and the restriction gene. We stumbled onto this story by accident but gained a rare and invaluable snapshot of how genetic protection against a lethal virus can evolve.

Few examples in the genetics or medical literature present such clear implications. The genes of living species are products of experience and natural history; and epidemics matter! Wild mice enjoy no medical surveillance or treatment to survive a fatal virus. The only defenses they have derive from intrinsic gene variation and the process of natural selection. A serendipitous aborted infection protected its recipient in a chance exploitation of retroviral interference, and the protection was passed on for centuries, increasing its numbers with each generation by conferring on its carriers a beneficial genetic vaccine against the plague.

AKV and *AKVR* are but a few of the endogenous virus vestiges that populate the genomes of mice today. Actually there are hundreds of copies of retroviral sequences in mice, some old, some young, and all genetic echoes of historic epidemics. We may never explain them all, but the resolution of this one is rather encouraging.

As we scroll through the genomic DNA sequences of other species of mammals, including our own, we encounter as many endogenous viral sequences as in mice. Most are dead vestiges of ancient diseases, but a few actively produce RNA copies and even synthesize proteins. Could some human genes have resistance functions like the *AKVR* gene, perhaps for HIV-1, the retrovirus that causes AIDS, for hepatitis viruses, or even for the deadly smallpox virus? No one is sure, but the exquisite tools of the human genome project may help reveal such connections. It is also possible that latent viral genes may awaken in the future to dispatch some new viral challenge.

Actively using genes to battle disease is a tantalizing prospect for twenty-first-century medicine. Beginning efforts of the futuristic gene therapy have been disappointing. We need to develop "smart vectors," molecular transport vehicles that can deliver genes to specific tissues, turn on their genetic passengers in just the right amount, in just the right place, and cause no damaging side effects. No one is sure how to do this yet, but the LC mice offer fresh hope because they did it all by themselves! The lessons learned from gene targeting in wild populations might just provide insightful guidance in human gene therapy trials.

The genomic secret of the Lake Casitas mouse came into sharp focus when the scientists of the Human Genome Project began to list millions, even billions, of DNA sequence letters from humans and other mammals on the Internet. The Human Genome Project is an international consortium of scientists, administrators, and funding institutions that share a single goal: to produce a full-length DNA sequence of our 3 billion nucleotides (the DNA letters of the human genome), which comprise our twenty-four different chromosomes. Alongside this effort are parallel attempts to sequence the genomes of other species, like the medically valuable mouse and rat.

One startling revelation of the first human genome sequences is the existence of thousands of stretches of DNA that are actually parts of retroviruses, like the *AKVR* restriction genes. In total, some 1% of human DNA is retroviral related. Most of these retrovirus genomes contain errors or spelling mistakes, which leads scientists to believe they are very old, disarmed, and descended from extremely distant ancestors. But some are full-length like AKV. Today's endogenous retroviruses have been passed down faithfully from parent to offspring, vestigial genomic footprints of fatal retroviral disease among our distant ancestors.

Charles Darwin once explained that in nature's quiet setting, "No fear is felt. Death is usually prompt and the happy, healthy, and vigorous survive and multiply." With our developing molecular and genetic technology, we actually possess the tools to reconstruct the details of evolutionary history. Unraveling these sagas may equip us to one day assure that history, at least that of deadly disease, does not repeat itself.

The LC mouse story came from a nontraditional source: illicit mice struggling under the squab dung in rural California. Perhaps the more usual approaches to science discovery, experiment design, and precise hypothesis testing are not the only ways to unveil biology's little secrets. Were there more genetic gold nuggets awaiting the attention of curious scientific sleuth-seekers in other wild species? Let us now turn to the fastest land animal on the planet—the beautiful and mysterious cheetahs of the African savannah.

Two

Tears of the Cheetah

SCIENTIFIC RESEARCH IS A CRAPSHOOT. YOU CAN RARELY predict the outcome and you never know what you might discover. Scientists try to focus their investigations on uncharted and mysterious venues because they are inquisitive and because they want to go where no one has gone before. Once in a while, a scientist stumbles onto results so unexpected that they fundamentally change accepted scientific beliefs. These discoveries can be driven by scientific brilliance, but more commonly, it is just luck. Perhaps most important is a willingness to recognize seemingly aberrant data as relevant, and to pursue its meaning relentlessly until the truth is uncovered and the skeptics are silenced.

The cheetah's genetic legacy makes this point. Well known as the world's fastest land animals, cheetahs display a potpourri of physiological adaptations that allow their magnificent, high-speed sprints on the African plains. Looking more like greyhounds than cats, cheetahs have elongated legs, slim aerodynamic skulls, and enlarged adrenals and heart muscles, plus semi-retractile claws that grip the earth like football cleats as they race after their prey at sixty miles per hour.

Cheetahs have mesmerized people throughout history. Regal potentates trained cheetahs as hunting companions, treasuring their predatory skills balanced by their timid acquiescence to captivity. King Tut's tomb was decorated with numerous cheetah statues and war shields sheathed by cheetah skins. William the Conqueror and the Mogul emperor Akbar kept cheetahs for sport hunting. In the chronicles of his Oriental adventures, Marco Polo reported that the Chinese emperor Kublai Khan kept over one thousand cheetahs in his summer palace at Karakoram.

Today cheetah populations are severely threatened, mostly by human depredation. The rise of agriculture has slowly but relentlessly consumed cheetah habitat throughout Asia and Africa. Cheetahs nearly went extinct in Asia in the 1940s and only a handful exist above sub-Saharan Africa, in Iran, today. Recent estimates place the entire African cheetah population at ten thousand to fifteen thousand. With more and more endangered species hemmed in by a human population approaching six billion, it is an all-too-familiar story.

The conservation movement grew out of a public concern stimulated when emotional treatises like *I Married Adventure* by Osa Johnson or *Out of Africa* by Karen Blixen forecasted the depletion of African wildlife. Rachel Carson's *Silent Spring* sounded the warning of a world inhabited by only a handful of domestic plant and animal species. Global resolve to halt widespread extinctions led zoos to revise their role from collectors and displayers of the world's diversity to partners in conservation. A first step was to build and to breed captive collections of endangered species like tigers, cheetahs, great apes, pandas, and condors as precious reserves or backups to fragile wild populations. Their hope was to learn more of the biology of endangered species and to develop management plans for protecting the species in their native habitats.

Cheetahs quickly proved to be difficult subjects. They are skittish, jumpy, even neurotic in captivity and they do not like to breed. Akbar's single cheetah litter in the sixteenth century was the only recorded successful breeding of captive cheetahs in over four thousand years of human-cheetah contact. The first captive cheetah birth in modern times occurred at the Philadelphia Zoo in 1956, but the cubs lived only three months. Since then, less than 17% of captive cheetah pairs have bred successfully in the handful of zoos attempting to procreate cheetahs. Among those that did produce cubs, mortality was 30–40%, higher than almost any other zoo animal. Even today, after considerable improvements in husbandry, the captive cheetah population is at risk of dying out because breeding cannot keep up with death rate. All the breeding tricks that worked with lions, tigers, and other large cats failed with cheetahs. Zoo directors and curators were puzzled and desperate for answers.

In the early 1970s, Frank Brand, an aggressive conservation-minded director of the Zoological Gardens of South Africa, joined with his close friend, chicken farmer Ann VanDyk, to set up the DeWildt cheetah breeding center near Pretoria. Brand and VanDyk had the same difficulties caring for cheetahs as the rest of the zoo world: jittery cheetahs, persistent breeding failures, and high cub mortality.

At an annual zoo directors' conference in 1980 in the United States, Brand discussed the cheetah's dilemma with the director of the Smithsonian Institution National Zoo, Ted Reed. Reed offered to send a medical team to South Africa to examine Brand's cheetahs to search for a cause of the cheetahs' breeding problem. Brand and VanDyk quickly agreed to a thorough medical evaluation of the DeWildt cheetahs.

Within a few months Mitch Bush, head veterinarian at the National Zoo, and David Wildt, a young reproductive physiologist working as a postdoctoral fellow in my laboratory at the National Cancer Institute, were on a plane bound for South Africa. Bush is a towering, bearded giant of a man with a strong interest and acumen in exotic animal veterinary medicine, particularly the rapidly improving field of anesthetic pharmacology. Wildt is a slight and modest midwestern farm boy, schooled in the reproductive physiology of barnyard animals. His boyish charm and polite shy demeanor mask a piercing curiosity and deep knowledge of all things reproductive. Bush and Wildt's expedition to the DeWildt cheetah breeding center outside Pretoria would ultimately change the way the conservation community viewed cheetahs forever.

VanDyk and Brand had accumulated about eighty cheetahs in the DeWildt compound and they offered them all to the American team. Bush gave each a veterinary exam and collected blood for hematological, hormone, antibody, and genetic tests. In addition, Wildt and Bush collected semen from fifteen adult males using an indelicate procedure developed for livestock termed "electroejaculation" (which is not as upsetting as it sounds). When Wildt inspected the sperm under a microscope, he was struck by what he saw. The sperm count in every cheetah male was very low, about one-tenth of what he was used to seeing in other cat species. More unusual was the rather high amount of malformed sperm: superlarge heads, tiny heads, bent or coiled tails. Such developmental abnormalities are found occasionally in other species like dogs or horses, but usually at frequencies less than 25%. Higher frequencies are typically found only in sterile males. Every cheetah Wildt tested had nearly 70% malformed sperm. No wonder the cheetahs had trouble breeding. But why were the cheetahs' sperm so aberrant?

Every clinician knows that vast medical information lurks in the blood. The precious fluid carries antibodies to bacteria, viruses, and the microbial agents encountered during a lifetime of infectious disease. Hormones reflect age, sex drive, pregnancy, stress, and composure, while steroids and fats are indicators of nutrition and organ function. Blood contains enzymes and DNA from damaged tissues and billions of copies of each individual's genetic code.

Bush and Wildt drew several tubes of blood from each of fifty cheetahs and laced it with heparin, an anticoagulant chemical that prevents clotting so the blood can be separated into its parts: red cells, white cells, and the plasma fluid. Stored in tiny 2 cc plastic tubes, the biospecimens were placed in larger steel cylinders lined with Styrofoam and supercooled to −196°C, −320°F, with liquid nitrogen. These containers were bound for my corner of the world, the Laboratory of Genomic Diversity of the National Cancer Institute, in Frederick, Maryland.

In those days my research group was small and focused on finding genetic differences among domestic cats that influence their susceptibility to the feline leukemia virus or FeLV, a retrovirus that induces blood cancers, leukemia, and lymphoma. David Wildt had explained to me the reproductive dilemma of cheetahs and his plan to sample large groups of African cheetahs. He wondered if there were ways to examine the genetic structure of the population and perhaps uncover something that might explain the cheetahs' reluctance to breed.

Back in the 1960s, scientists developed a procedure called gel electrophoresis to separate blood proteins based on their electrical charge. Blood enzymes, proteins that catalyze or facilitate a chemical change (e.g., break down a sugar molecule to generate energy), can be subjected to electrophoresis and stained with colored dyes that identify particular enzymes. For most species, common variants in the genetic regions responsible for encoding these enzymes cause mobility shifts that can be monitored easily on an electrophoresis gel. Different electrophoretic migration patterns indicate subtle differences in the DNA sequence of the gene that coded the individual enzymes. The variants of a single gene found in a population are called alleles, so the enzyme's genetic differences were called allozymes, short for *allelic enzymes*.

Population geneticists have long been interested in estimating how much genetic variation between individuals at the DNA level is normal, or, for that matter, abnormal. This is no small task. The amount of DNA present in a single sperm or egg from a typical mammal (human, mice, cats, dogs) is about three billion nucleotide letters. These are all arranged in a linear array packaged in twenty to thirty unique chromosomes. Different species have slightly different chromosome numbers; for example, humans have 23 pairs, mice have 20 pairs, and cats have 19 pairs. The vast majority of chromosomal DNA in any mammal is noncoding "spacer" DNA, but around sixty million nucleotide letters comprise the coding material for approximately thirty-five thousand genes. Enzymes are typical gene products, so sampling a group of them by electrophoresis from several individuals of a population would produce a rough estimate of the overall amount of genetic variation present in the population.

One thing that had changed in the decade since my graduate school days was the sheer volume of allozyme population estimates of genetic variability. By the time the cheetah blood samples arrived at our laboratory, close to one thousand population surveys of allozyme genetic variation had been reported on all sorts of species, from flies to plants to bacteria to birds to cats. Nearly every surveyed species produced the same conclusion. Between 20% and 50% of sampled allozyme genes had discernible genetic variation. This means that a typical individual is heterozygous (i.e., has two different alleles at the same enzyme gene inherited from different parents) at between 5% and 20% of his allozyme genes.

In our ongoing effort to develop genetic tools to study cats, my students and I had already developed testing procedures for some sixty cat allozyme genes. As soon as we could, we thawed the frozen cheetah blood tubes, prepared cell extracts rich in cheetah enzymes, and began a population genetic survey of the cheetahs from DeWildt. We expected to find a modest level of genetic variability and to report that cheetahs had normal levels of genetic diversity. But that is not what we found.

Janice Martenson was a brilliant young technician recruit in the lab, whom I had snared from a career as a lab-ware salesperson with the promise of an exciting future in the relatively new field of molecular genetics. Already well trained as a virologist, she picked up the new molecular technologies quickly, including the allozyme methods. Jan began scrolling through the cheetahs' blood samples with the allozyme assays. Several weeks into her work, she commented irreverently, "These cheetahs sure are boring. I can't find any variation!" And so it went. Enzyme after enzyme, every sample was identical. We stopped at fifty-two allozyme genes. Every cheetah was identical across the board.

Years later I would joke to Mitch Bush, "You guys didn't really collect fifty cheetahs, did you? What you actually did was to collect one cheetah's blood, then split the blood into fifty separate tubes, right?" The results were that hard to believe. Every cheetah was genetically identical to all the others. Their genes had the look of deliberately inbred laboratory strains of mice or rats.

To create such inbred rodent strains, siblings are mated together by breeders for ten to twenty generations straight. This process causes a wholesale shedding of common genetic DNA variation and results in near zero DNA variants. But cheetahs are wild animals; they mate at random, and extensive field observations of cheetahs made no mention of any incest, let alone serial incest. We thought something must be wrong. Maybe allozymes were deceiving. Maybe all cat species have variation lower than other animals. Perhaps we were simply at the low end of a statistical distribution of one thousand population surveys. And what about wild as opposed to captive cheetahs or the cheetahs from East Africa, which were supposedly a separate sub-species? How would they look? We needed to collect more data.

I assigned a bright new graduate student, Andrea Newman, to test the same allozyme genes on eight other cat species. Using blood samples we had on hand thanks to Bush and Wildt's numerous sperm collection trips over the years, Newman found considerable genetic variation (10–30% of allozyme genes had multiple alleles) in lions, tigers, leopards, caracals, servals, ocelots, and even domestic cats. We then looked at a different group of the cheetahs' gene products, 155 proteins measured in tissue culture cells, originally grown by culturing cheetahs' skin biopsies in nutrient medium. The cheetahs' skin cell proteins also showed dramatic, but not complete, reduction in variation compared to other cat or mammal species. At least there was some variation, but precious little.

By now, we knew our results were potentially very important, but I was still a bit uneasy about the extremity of the finding. Only one other allozyme survey of any wild population had been so low. A genetic sampling of the giant northern elephant seals, best known for their big Jimmy Durante noses and ability to dive awesome depths, had also revealed 0.0% allozyme variation, but that study was based on only twenty-four enzyme genes. The seals had been decimated by overhunting during the late eighteenth century. In fact, they were presumed extinct as late as 1912 when a group of eight seals was spotted on Guadalupe Island in the South Pacific by a Smithsonian expedition, which shot seven of them! Fortunately they missed one and failed to spot a few others. In 1920, the few remaining seals were afforded legal protection as endangered species off the California and Mexican coast, leading to a gradual recovery. The population grew to a high of 120,000 by 1960, forty years after the U.S. and Mexican governments agreed to protect them. The genetic depletion in northern elephant seals is thought to be a consequence of their nineteenth-century brush with extinction, exacerbated by their strange harem-based breeding system. Usually dominant alpha males inseminate hundreds of females while envious bachelors look on in the hope that they might someday be promoted to alpha male.

But cheetahs were never systematically slaughtered for their skins. Their range was vast, covering most of sub-Saharan Africa for thousands of years, and their elusive nature protected them effectively from hunters. It was time to examine more genes to be certain our sampling of the genome was representative.

We turned our attention to a gene group that we knew a lot about called the major histocompatibility complex, or MHC. The MHC is a cluster of some 225 genes that occur together along a short chromosomal segment in the DNA of humans, mice, cats, and other mammals. About a dozen of the MHC genes encode proteins that coat the surface of cells where they engulf small peptides (short stretches of amino acids) from invading viruses as a prelude to immune-mediated demolition. Most of the MHC genes are extremely variable; some have over two hundred different alleles in outbred populations of mammalian species.

The reason for the extensive gene variation was a mystery to immunologists for most of the twentieth century until two Australian researchers, Rolf Zinkernagel and Peter Doherty, proved that the MHC proteins act as a conduit for serving viral proteins of infected cells to a vigilant immune system. Blood lymphocytes, or white cells, recognize the MHC-bound peptides as enemy invaders and quickly destroy the cell and its virus. The wide variety of genetic types available at multiple MHC genes offers individuals a vast repertoire of foreign peptide recognition specificity necessary to battle the plethora of microbes the immune system encounters in a lifetime. The revelation that mammalian MHC provided the basis for immune recognition and defense against invading microbes was so fundamental that its discoverers were awarded the Nobel Prize for Medicine and Physiology in 1995, thirty years after their original proposition.

That the MHC was extremely variable between individuals was well known to organ transplant surgeons because the MHC gene products, proteins expressed on nearly all cell surfaces, are themselves antigens—that is, proteins that can elicit a powerful immune reaction. Kidneys or livers transplanted between unrelated people rapidly lead to graft rejection, an attack by the immune system on foreign MHC antigens of the implanted organs. A tissue match between unrelated people is very rare, less than one in ten thousand. This is why patients in need of kidney, liver, and heart transplants wait so long, and why local motor vehicle associations try to get us all to become organ donors.

A clearer understanding of the genetic basis of graft rejection came from mouse geneticists who exchanged surgical skin grafts among laboratory mouse strains. When small pieces of skin were removed from an individual and replaced in the dime-sized original graft beds of the same animal, they would heal in a few weeks. However, when pieces of skin from unrelated individuals were sutured into a graft bed, the immune system recognized the foreign MHC antigens as being "nonself" and launched an attack. The skin darkened and became hard as immune cells invaded the genetically foreign tissue and formed a scab, due to acute immune cell infiltration of the graft. But when grafts were exchanged between mice from the same inbred strain, ones with identical MHC genes, the grafts healed and were accepted as if they were "self."

Cheryl Winkler loved the MHC. As a graduate student she set up a cat colony and learned to perform skin grafts from burn unit surgeons at Children's Hospital in Washington, D.C. She exchanged skin grafts between the colony cats skillfully, using antiseptic surgical procedures to characterize the MHC of the cat in detail. For every cat, unrelated grafts were rejected twelve to fourteen days after surgery, an indication that their immune systems recognized and rejected foreign skin and unrelated MHC antigens.

When the cheetahs' remarkable genetic uniformity was revealed by Janice Martenson's allozyme screen, we wondered if the granddaddy of genetic diversity, the MHC, was similarly homogenized. Cheryl and I trained Mitch Bush to perform the delicate skin transplant surgery and dispatched him to South Africa to graft six unrelated cheetahs. If their genes were as monotonous as we thought, their immune systems would not reject grafts from other unrelated individuals. Bush and Dr. Woody Meltzer, head veterinarian at the Pretoria Zoo, exchanged skin grafts between six captive cheetahs, two brothers and four unrelated cheetahs selected for the test. Each received a control autograft, skin from the same animal, and an allograft, a piece from an unrelated cheetah. On day 14 after the graft, back home in the United States, Bush and I telephoned Meltzer long-distance to ask him how the cheetahs were doing. They were recovered, lively, and doing well.

"What about the grafts?" we demanded. Meltzer said three words: "They're doing fine!" He meant they were healing, both autografts and allografts. No difference. Two weeks after that telephone conversation, now four weeks after surgery, the grafts were still recovering. By day 45, hair was growing out and the cheetah spots were appearing. All six cheetahs had accepted grafts from other cheetahs; each healed as if they were genetically identical.

I was still incredulous. How could this be? Cheetahs were wild animals, not some inbred strain. Meltzer was a brilliant veterinarian, but maybe he wanted to see this result too much? We needed more confirmation. We had to be certain.

By this time we were aware of a successful breeding collection of about sixty cheetahs living in a rural drive-through wildlife park in Winston, Oregon, called Wildlife Safari. Founded in 1972 by a Los Angeles real estate developer and big game hunter turned conservationist, Frank Hart, Wildlife Safari offered the perfect place to repeat the skin grafts. Hart had two enthusiastic and tireless lieutenants: Laurie Marker, studbook keeper for cheetahs throughout the United States, who combines an encyclopedic knowledge of cheetah fact and lore with a determined conservation ethic that exceeded any I have encountered before or since, and her sidekick, Dr. Melody Roelke, the park veterinarian who made it her mission to understand everything about cheetah veterinary medicine, reproduction, and infectious disease. Marker and Roelke were terrified, but also fascinated, by the implications of our genetic results. They persuaded Hart to allow our team to visit and put their cheetahs to a test of science, repeating the skin graft experiment that Bush and Meltzer had performed in South Africa.

Bush, Wildt, and I flew to Oregon with our surgical supplies. On Valentine's Day, 1983, we surgically grafted eight cheetahs, two sisters and six unrelated cats, in the veterinary clinic at Wildlife Safari. Each cheetah received an autograft (self) and an allograft (nonself), and two of the cheetahs also received a skin graft from Roelke's pet domestic cat named Heidi. This domestic cat piece, called a xenograft since it was from a foreign species, provided what scientists call a positive control; that is, it would show that the cheetah's immune system was in fact functional and could recognize and reject foreign antigens. Typically in Cheryl's graft experiments with domestic cats, allografts and xenographs would be rejected in ten to twelve days. So on February 28, fourteen days after the grafting, Cheryl and Melody carefully removed the bandages from the eight cheetahs.

The grafts looked good. Eight of eight autografts were accepted. Eight of eight allografts were also accepted and healing. But, as we hoped, the two xenografts from Heidi showed acute rejection. Cheryl and Melody were not completely convinced and wanted to observe the grafts a little longer. News of the exciting results came to me burdened with a growing sociopolitical situation, not uncommon with research on high visibility species. Some of the Winston citizens and supporters of the park were resentful of "government agents descending on Winston to molest their cheetahs as if they were laboratory rats." The director, Frank Hart, was queasy and decided he should terminate the experiment. Cheryl and Melody spent hours trying to dissuade him. When I heard of the flap, I immediately flew to Oregon, where Frank met my plane. After several hours, many beers, and a few cigars in a cowboy saloon along the state highway in Winston, Frank made a courageous decision, one that would have implications for cheetahs and species conservation for a generation. He urged me to continue to monitor the cheetahs and to follow through with the experiment.

Within weeks, the results became crystal clear. The allografts all survived through the acute phase, and the xenografts were immunologically rejected right on schedule. Three of the allografts showed signs of rejection in late March, fifty days later, but this was because of much slower non-MHC antigen differences. The cheetah's immune system was functional, but it did not recognize MHC differences among unrelated individuals. The genetic status of the cheetah was dramatically homogenized as if there had been some historic purging of genomic diversity. Such a degree of immunological identity was incredible and had only been seen before among identical twins or in the very special contrived cases of inbred strains of mice or livestock.

The evidence of cheetah genetic monotony would only grow. Bob Wayne, a talented postdoctoral fellow in our lab, examined cranial measurements and the bilateral symmetry of cheetah skulls. Although no one is certain why, in most livestock, asymmetry in skeletal characters (the difference between right and left measures of a trait) increases with inbreeding. Bob measured sixteen bilateral traits in thirty-three cheetah skulls held in natural history museums in Washington, Chicago, and New York. The study was not perfect because several of the skulls were incomplete due to a bullet hole in the skull! Nonetheless, in nearly every case, cheetah skulls were more asymmetric compared to the skulls of leopards, ocelots, or margays. When I explained these skull results in a television interview, the correspondent asked, "Dr. O'Brien, are you telling me that these cheetahs are lopsided?" Not exactly, but the cheetahs certainly looked very inbred.

The implications of each of these studies were beginning to coalesce. No matter how we looked, cheetahs appeared genetically very, very similar. Yet, by 1985 all the cheetahs we had examined were from southern African locales: Kruger Park in South Africa, the Kalahari Desert in Botswana, and private farms in northern Namibia. Could it be that the cheetahs from East Africa, a distinct subspecies of cheetah, were different? The one clue we had came from Bob Wayne's lopsided cheetah skulls. Most of these were collected in East African game parks around the turn of the century by Teddy Roosevelt's hunting safaris. And these skulls showed evidence of genetic homogenization.

To be certain, we had to look closely at the genes of cheetahs from East Africa. It took a year to organize a collection trip. The National Geographic Society offered to fund our expedition and we enlisted two experienced cheetah watchers, Swiss wildlife photographer Karl Ammann and British ecologist Tim Caro, to help us locate wild cheetahs on the Serengeti Plains in Tanzania and Kenya. Paleontologist Richard Leakey, director of the National Museums of Kenya, threw his support behind our quest, greasing the political skids for what had never been done before: sampling free-ranging cheetahs for reproductive and genetic status. Bush, Wildt, and I arrived in East Africa in June 1985 for a six-week expedition. We brought eighteen crates of supplies—generators, centrifuges, electroejaculator probes, plasma tubes, and liquid nitrogen—a virtual mobile lab. We purchased a very used 1976 Toyota Land Cruiser from a Kenyan missionary, dubbed it NOAH, and set out with Ammann, Caro, and Leakey's lieutenant Issa Agundey for Seronera, a remote village within the Serengeti where wildlife researchers congregate to study the vast but vanishing East African ecosystem.

Cheetahs hunt at dawn and dusk, so we were out before sunrise each morning, peering through binoculars, hoping to spot cheetahs sitting atop termite mounds. Thanks to Tim Caro and Karl Ammann we found dozens of cheetahs and were able to dart, anesthetize, and collect samples from thirty free-ranging animals. Wildt collected semen from ten males and every one showed low sperm counts and highly malformed spermatozoa akin to our South African cheetahs. Their blood, skin, hair, and even fecal material were processed, frozen, and shipped as excess baggage on our flight back to the United States. We wasted no time, working quickly to determine their genetic status. Of the allozyme genes, 96% were uniform and identical to South African gene forms, while 4% of the genes displayed a little variation. This was more gene variability than was seen in the southern animals, but just barely. We found what we had feared, that both remaining cheetah subspecies, *Acinonyx jubatus jubatus* (from southern Africa) and *Acinonyx jubatus raineyi* (from eastern central Africa), had 90–99% less genomic diversity than other cats and reproductive characteristics that were noticeably impaired. The entire species had somehow lost a huge portion of its original genetic variability. But how did this happen?

The truth is that we are not really sure how the cheetah lost its diversity, but we can make some educated guesses. The simplest explanation would be a scenario similar to that of the northern elephant seals, a historic near-extinction event followed by inbreeding. Mammals have hardwired genetic instincts to avoid mating with close relatives. Inbreeding avoidance behaviors have been well documented by ecologists studying cheetahs, lions, and other big cat species. Inbreeding produces congenital abnormalities by bringing together rare recessive alleles—those that have been damaged by mutation, a spontaneous error of a gene's replication. Recessive alleles require two doses, one from each parent, to produce a defect; in contrast, one dose of a recessive allele plus a normal allele of the same gene masks the problem. If there are normally two to three million "DNA letter spelling" differences among individuals (which there are in outbred species) and most are recessive and ill-suited for optimal development, then breeding among close relatives will lead to harmful recessive allele collision or co-occurrence in offspring. Increased congenital abnormalities following incestuous matings, termed "inbreeding depression," are well known in livestock. The phenomenon occupies an entire chapter in *On the Origin of Species* by Charles Darwin. It would seem that the cheetahs' elevated cub mortality, reduced fecundity, bad sperm, and reproductive and congenital abnormalities were the consequence of forced inbreeding in their past.

The best explanation, one that squares with the cumulative data about cheetahs, involves a historic brush with extinction. A long-forgotten cataclysm, whatever it was, probably reduced cheetahs' ancestors to a very small number. In effect the cheetah species passed through a "population bottleneck." Their numbers dropped so low that they abandoned instinctive avoidance of incest and mated with close relatives because there were few other mate choices available. Inbreeding must have persisted across several generations to produce the level of genetic depletion seen in modern cheetahs.

The cheetahs' near extinction is now believed to have occurred about twelve thousand years ago, toward the end of the Pleistocene, a geologic period that brought the most recent ice age to the Northern Hemisphere. Before then, at least four species of cheetahs ranged throughout Europe, Asia, Africa, and North America. But after the last glacier retreat, cheetahs abruptly disappeared from all but Africa, India, and the Middle East. At about the same time that the cheetahs' global range contracted, the world also saw the loss of three-quarters of the large mammals living on those same continents. This global extinction was abrupt and the most extensive species loss that occurred in the seventy-million-year history of mammals. The event eliminated some thirty-five to forty large animal species in a very short time, including giant ground sloths, mastodons, saber-toothed tigers, and American lions. No one is certain of the cause. Some say it was climatic changes; others blame human hunting pressures, or even a devastating infectious disease of the big mammals or of their prey. Whatever the cause, the cheetahs escaped extinction at around the same time by establishing a refuge in Africa.

The cheetahs' genes actually confirm this time frame. The amount of population genetic variation that will accumulate in a species after a population bottleneck is a function of time. The longer it has been since the crash, the more DNA variation becomes reconstituted. In a computation reminiscent of monitoring the radioactive carbon decay to date fossils, geneticists can assess the time elapsed since a population bottleneck by calculating the time it takes to produce the observed quantity of new variation in rapidly mutating genome families.

We chose to examine three DNA regions that we knew would accumulate new DNA variants a hundred times faster than the monotonous allozyme and MHC genes. First we examined the relatively short sixteen-thousand-nucleotide-letter circular chromosome found in cell organelles called mitochondria. Mitochondria serve as a cell's powerhouse where energy molecules are manufactured by combining atmospheric oxygen with a breakdown product of carbohydrate nutrients. All plant and animal mitochondria are themselves descended from a 600-million-year-old bacterial infection of early one-cell organisms. Today's mitochondrial DNA carry the remnants of that primitive invading bacteria's genes and use them to make energy-rich molecules. Because the mitochondria's chromosome is found outside the nucleus in the cell cytoplasm, it doesn't benefit from the DNA repair enzymes found in the nucleus and so is subject to a very high incidence of uncorrected mutation—DNA spelling mistakes that occur during replication of DNA. The new mutations accumulate rapidly in the generations following a population bottleneck like the cheetah experienced. Two additional rapidly evolving nuclear DNA categories that we examined are microsatellites and minisatellites. Microsatellites are short, repetitive sequences of two to five nucleotide letters, while minisatellites have longer repeat sequences of twenty to sixty nucleotide letters. For complex reasons, both micro- and minisatellites are prone to high mutation rates and accumulate scores of new alleles after a population bottleneck.

When we measured the amount of the cheetahs' diversity in the three DNA marker categories, we discovered that each group showed a great deal of new variants, enough to provide an estimate of the time elapsed since the original bottleneck. The calculation for the cheetah gene segments computes back to ten thousand to twelve thousand years ago, smack on the date of the great mammal extinction events at the end of the Pleistocene. So whatever had decimated the saber-tooths, mastodons, and giant sloths had nearly claimed the cheetahs as well. That narrow escape from extinction left its mark of genetic uniformity on the survivors, only to be discovered ten millennia later by our research team. The puzzle was slowly coming into clarity.

Consider that time long ago when the cheetah roamed across the globe alongside the prosperous herbivores and carnivores of its day. Imagine a young pregnant female somewhere in southern Europe who climbed into a warm cave to slumber through a harsh winter. When she and her cubs crept out in the springtime, they met a different world, one in which the cheetahs and great predators of the region were gone, victims of a global holocaust.

Multiply such an eerie scenario over time and space, and we have a recipe for extinction, or in rare cases like that cheetah matriarch, a population bottleneck. The founder of a new race, a reservoir of the accumulated gene adaptation that created such a magnificent evolutionary masterpiece, would mate with her sons and continue the legacy of her species' recovery. My reverie conjures a vision of the tears on that cheetah-mom that would render an indelible tear-stripe below the eyes of every cheetah from that moment onward.

Explorers and scientists alike occasionally experience an epiphany that instantaneously fuses countless small details into a dramatically focused reality. Such a drama unfolded at the Oregon cheetah breeding facility during the days of our skin graft surgery experiments. We were blind to the big picture at the time.

Wildlife Safari started its modestly successful cheetah propagation program with 6 animals in 1973 and had a population of 60 by 1982. In October of that year, two cheetahs, Sabu and Toma, were brought to the wildlife park from the Sacramento Zoo. Within weeks of their arrival both animals developed fevers, jaundice, severe tremors, and diarrhea. In spite of heroic efforts by Dr. Melody Roelke, the park's veterinarian, both cheetahs perished.

Sabu and Toma were infected with a nasty virus called "feline infectious peritonitis virus" (FIPV), which had been identified in domestic cats a decade earlier. In house cats FIPV stimulates a strong immune response that results in a gradual accumulation of immune protein chunks in the cat's belly or peritoneum. In some cases, the complexes become so dense they strangle the kidney, liver, and eventually all internal organs, leading to a rapid and excruciating death. This is how the two cheetahs had died in Winston. The virus actually co-opts the cat's own immune system, transforming it to an agent of the pitiful animal's destruction.

Several outbreaks of FIPV had been described in experimental domestic cat colonies in veterinary schools and catteries (a cattery is a private household where dozens of cats are raised and nurtured for the joy of their company). Mercifully, the mortality in those episodes was low, typically 2–5%, and seldom over 10%. The reason for the low mortality was thought to be the innate genetic diversity in the domestic cats' immune response. Most cats have an immune system that stems the fatal disease progression, while a few genetically deficient individuals do not.

For eons, infectious disease agents and their hosts have engaged in an evolutionary arms race, pitting virus genes against host immunity genes. Millions of years of vertebrate evolution has produced complex and elaborate immune defense mechanisms. A critical part of that defense is the genetic diversity of the species itself. When an emerging new virus or parasite encounters a new population, immune defenses will thwart it, but the virus can evolve to circumvent immune defenses, causing individuals to succumb. Yet a variant virus that overcomes the immune defenses of some individuals will still be dispatched by others because every animal is genetically distinctive. An outbred population is a formidable moving target for a virus and that is why house cats have low population mortality when they are hit with FIPV.

Melody Roelke hates not knowing what is happening. The Winston veterinarian is a compulsive worrier, observer, and collector. When she first signed on at Wildlife Safari, she banked away frozen blood samples from every animal for clinical monitoring. Years later, when she diagnosed Toma and Sabu with FIPV, she worried about the other cheetahs, so she had all their old plasma tested for antibodies to FIPV, a sign of infection. The result: Every cheetah's serum sample collected before the arrival of Toma and Sabu was negative for FIPV antibody. So no FIPV had appeared or infected the cheetahs before the arrival of Toma and Sabu. The two were the Winston cheetahs' patient zero. But then things began to unravel.

By the end of 1982, FIPV antibodies had developed in every animal in the park. Then the plague began. Wobbly, jaundiced cheetahs with high fevers started to collapse. By spring of 1983, every single cheetah had been infected and showed symptoms of the disease. Overworked and overstressed, Melody Roelke and Laurie Marker attended hopelessly over the next several years, watching the animals die. The total death count was massive: 60% dead, 85% of the cubs. Roelke and Marker were devastated. They had witnessed and documented the worst FIPV mortality in any cat ever recorded, and it hurt to watch.

The Winston cheetah outbreak was a tragedy, but it taught the world an important lesson. FIPV was thought to be capable of a few mortalities, but its furor had been mitigated until now. The virus had overcome the defenses of Toma and Sabu, and the cheetahs it encountered at Wildlife Safari were virtually immunological clones. The disease spread like wildfire. The cheetahs' immune systems all offered the same fertile ground for the raging epidemic. An FIPV strain that had dodged the immune defenses of a single cheetah easily outpaced the disease immunity of them all.

When we announced our portrait of the cheetah's genetic history at a wildlife conference, the conservation community was stunned. First, it meant there were hidden perils, hangovers from past events beneath the surface of at least one endangered species, that science had overlooked. We would never have seen it by traditional ecological monitoring—peering through a pair of binoculars from atop a Land Rover. The irony that technology, itself a principal cause of species endangerment, might actually assist in reversing an extinction vortex had not escaped us.

The second message was more troubling. When a threatened species is fortunate enough to dodge an extinction crisis, their future is doubly precarious. Even if the census rises back to hundreds of thousands, as it did for cheetahs and northern elephant seals, the recovery may depend on close inbreeding, shedding gene diversity, with devastating consequences.

One glaring outcome is an unpredictable array of congenital impairments that can cause fetal/infant abnormalities as well as defects in reproduction. Both were seen in cheetahs from South Africa, in cheetahs from East Africa, in free-ranging cheetahs, and in captive cheetahs. The second curse is less apparent but more insidious: the cryptic depletion of immune gene breadth. A species that loses its immune diversity carries a genetic axe suspended over its head, awaiting the next emerging fatal virus, bacteria, or parasite to appear. Could similar genetic losses have been the penultimate step in the extinction of some or many species long gone? Some scientists believe that infectious disease outbreaks play as large a role as ecological pressures (e.g., climate, predators, prey, etc.) on the demise of species. If that is true, the cheetah's lesson is particularly disturbing.

Is there any silver lining to all this? For the cheetahs, perhaps. As desperate as the cheetah's genetic situation appears, it is probably a mistake to consider them doomed. The most serious damage to a population after a bottleneck occurs immediately. In fact, most populations that drop to a few individuals simply die off.

Cheetahs did not die off. They have survived for twelve thousand years and increased in number to hundreds of thousands, their count as recently as a few hundred years ago. The cheetah's genetic problems, however real, have not limited their population growth appreciably. And their solitary nature, while demanding vast spans of habitat, probably retards the spread of microbial diseases by limiting individual contact. The cheetah's biggest problem this century is human-driven habitat loss. Hope for survival is less dependent on the past than the future. If conservation initiatives succeed in a resolve to protect the cheetah and its habitat, my bet is that the cheetah will prevail.

The cheetah saga, particularly the graphic cost of population and genetic reduction, has implications that extend beyond the cheetah and even beyond conservation. The cheetah reminded the medical community of the critical role that genes involved with immune defenses, such as the supervariable MHC, are playing in maintaining human health. This led me to ponder the multitude of debilitating human diseases that fill our hospitals for which we have no cures and only symptomatic treatment: cancer, arthritis, over two thousand hereditary maladies, and fatal microbes like HIV, Ebola, and hepatitis. Could other animals reveal natural genetic defenses to these same diseases? Without such defenses, they will go extinct. These species have no health insurance, no HMOs, no emergency rooms—only natural selection.

How many other natural history stories like the cheetah's legacy lie in wait of discovery? Did other species have narrow escapes from extinction? Could there be additional equally tantalizing examples of near extinction and genetic accommodation out there? We set out to search for answers and lessons from nature's vast experience.

Three

Prides and Prejudice

MAYBE I WAS NOT PREPARED FOR THE EXTRAORDINARY reaction to our announcement of the cheetah's genetic history. We published the results in two *Science* articles in 1983 and 1985, followed by an overview by Bush, Wildt, and me in *Scientific American*. The popular media picked up the story and the work was featured in TV specials, magazine articles, and radio interviews. I received numerous lecture invitations and the cheetah's genetic secret became the buzz of the wildlife conservation crowd. But not everyone was singing our praises.

We had to contend with serious skepticism. Ecologists who worried that molecular genetics was receiving too much attention grumbled that cheetahs had done fine for thousands of years prior to human destruction of their habitat. Some critics wondered if the physiological problems we observed in captive animals reflected not genetic strain but simple stress induced by limiting a species whose ancestors roamed tens of thousands of hectares to cramped quarters. Furthermore, since all cheetahs were highly inbred and displayed reproductive or congenital problems, it was difficult to prove cause and effect. Certain pundits at my home institution, the National Institutes of Health, asked why they were supporting research on cheetahs in the first place. Cheetahs were hardly traditional subjects for medical research. What relevance did they have to curing cancer or other human diseases?

In spite of the doubting chorus, I was convinced we were onto something. Our gathering of experts from multiple biomedical disciplines around the cheetah had produced unprecedented insight into the perils faced by these majestic creatures. David Wildt brought to our team a seasoned experience in all the details of animal reproduction, from hormones to sperm development to assisted reproduction technologies. Mitch Bush had spent his career testing, tuning, and assessing pharmaceuticals for optimal animal handling. Recent developments in veterinary medicine offered remarkably safe and effective anesthesia for wild animals, particularly for big cats. And our timing was fortuitous: Molecular geneticists had just developed the tools for precise population diversity assessment. We stood in awe of the mystery we had unlocked.

My telephone started ringing off the hook. Field biologists studying large and charismatic animals wanted to know if their own species had genetic problems. I listened carefully to stories of koalas in Australia, giant pandas in China, black-footed ferrets in the Midwest, elephants, rhinos, and leopards in Africa, and orangutans in Asia—all threatened or endangered species attended by packs of worried field biologists. If cheetahs paid a price for their brush with extinction, did these species suffer the same?

Craig Packer's story drew me in. His study species was arguably the most charismatic of all: African lions. A cat, our specialty, and a species with considerable research inquiry already, the lion offered the perfect next step for our investigations. There are an estimated thirty thousand to a hundred thousand wild lions surviving in the game parks of eastern and southern Africa, where they are a popular tourist attraction.

Like cheetahs, lions had been treasured by kings, pharaohs, and monarchs for centuries. Paintings and sculptures from Asia to Egypt to Europe celebrated lions as the ultimate symbol of strength and power. Julius Caesar sacrificed four hundred lions when he opened his forum, and the Egyptian pharaoh Ramses II was accompanied by lions in his fiercest battles. Our adulation continues today from the roaring MGM logo to Broadway's smash hit *The Lion King*.

Craig Packer was a tall, wiry, bearded Texan with a keen intellect and an acerbic wit. He knew that the outstanding questions about lion behavior and survival far outnumbered the answers. He and his research partner and wife, Ann Pusey, who met when both were studying Jane Goodall's chimpanzees in Gombe, Tanzania, had taken their young family to the Serengeti National Park to manage the longest running lion ecology study ever. The pair shared a single assistant professor post at the University of Minnesota, but their hearts and minds were in the majestic East African plains.

The Serengeti-Mara ecosystem of Tanzania is a vast savannah plain of twenty-five thousand square kilometers (about the size of the state of Connecticut) defined by the migration patterns of twenty-eight herbivore species. Tony Sinclair, in his two-volume compilation of Serengeti ecological research, notes that the region offers a huge natural laboratory with a four-million-year history. Unlike other similar habitats, the Serengeti still teems with wildlife and remains unblemished by the settlements of modern humans. The unparalleled species diversity of the Serengeti became apparent only in 1957 when Bernhard and Michael Grzimek documented the scope of the migrations in detail. The keystone species, the wildebeest or gnu, numbers over 1.3 million today. Other species including 240,000 zebras, 444,000 Thomson gazelles, and associated predators comprise the richest density of large fauna on earth. In the 1880s, however, populations of wildebeest, cape buffalo, and several other ungulate species were decimated by an outbreak of rinderpest, a fatal measles-family virus that spread from domesticated Indian zebu cattle. The virus ravaged the Serengeti's hoofed species for nearly a century until a vaccine program in local cattle eliminated the disease in the 1960s.

The large migrating Serengeti herds have provided abundant prey for several carnivore species, including hyenas, leopards, wild dogs, lions, and cheetahs. In the late 1960s George and Kay Schaller undertook their classic study of lion behavior in the Serengeti. They observed that unlike all other cat species, lions are social, even communal. Lions live in close female-dominated groups called prides that include sisters, mothers, aunts, and cubs. Each female group defends a large territory and mates with a resident coalition of males who have won a competitive power struggle for access to the pride. Few resident coalitions last longer than three years because wandering nomadic males continually challenge the residents in hopes of a takeover.

These takeovers can be rather brutal rituals. The physically dominant male coalition threatens, fights, injures, and, in some instances, kills the losing males. Marauding males then methodically kill young cubs produced by the previous fathers. The mothers defend the cubs, but inevitably fail. Remarkably, within a few days of the infanticide, the widowed and now cubless adult females enter estrus and begin day-long mating rituals with the new resident males. The takeover actually appears to trigger the onset of estrus.

Craig and Ann began their tenure as curators of the Serengeti lion study in 1978 following John Elliot (1970–72), Bryan Bertram (1972–75), and Jeannette Hanby and David Bygott (1975–78), who had extended behavior observations initiated by the Schallers. Craig and Ann's curiosity, experimental acumen, and innovative approaches to the myriad questions surrounding the lion's behavior are now considered groundbreaking by their peers.

During the many hours Craig and I were to spend combing the Serengeti for lions, he explained the potential advantages and evolutionary adaptations that the lion pride organization allowed. A pride's females enter estrus at the same time and give birth synchronously. They communally nurse their sisters' cubs along with their own and defend their territory as a group. Such a strategy would improve cub-rearing and ensure against the loss of individual cubs. The females also take on the lion's share of hunting, and the group strategy has the advantage of increased success (less than 20% of lion hunts succeed) and also minimizes the time required to defend a kill from hyenas and other scavengers. The apparent advantage of cooperative hunting can backfire, however, since a kill has to be shared among multiple diners.

Lion cooperation seems somewhat "hardwired" or genetically programmed. Juvenile males remain with their pride for up to two and a half years before dispersing, far longer than other feline species with a more solitary lifestyle. Mating is also a group event, with multiple males serially copulating with females in estrus. It is fascinating that the observing males do not seem to object; they simply wait for their turn.

Lion habitat has been gradually disappearing for decades and the Packers wondered whether the Serengeti lion might be at risk for small population effects. There were plenty of lions around—Craig estimated around three thousand in the Serengeti-Mara ecosystem— yet a century of rinderpest-inflicted population crashes of wildebeest and buffalo prey could easily have caused unobserved lion population bottlenecks well before the Schallers' observations began in the 1960s. He asked me to look at the genetic status of the Serengeti lions.

The Packers also brought another interesting situation to my attention. About forty miles southeast of the Serengeti Preserve sits a long-extinct volcanic caldera, the Ngorongoro Crater. Two-thousand-foot-high mountainous walls surround a 250-square-kilometer crater floor (one hundred square miles—about the size of the District of Columbia), which is covered with dense vegetation. The area receives more water than the arid Serengeti and enjoys a rich diversity of East African wildlife: wildebeest, gazelles, hyenas, lions, cheetahs, even a few rhinos. A population of about forty adult lions and as many subadults and cubs were feasting on a lavish herbivore buffet.

For lions, the Ngorongoro Crater is like an island, protected by mountainous walls from immigration. The few lions from the neighboring Serengeti that try to migrate in are quickly dispatched by the resident males guarding the territory. Crater lions are well nourished, procreative, and healthy. For lions, the Ngorongoro Crater is paradise; but it was not always that way.

The spring of 1962 was particularly wet in Ngorongoro, leading to the unprecedented proliferation of a bloodsucking fly, *Stomoxys calcitrans*. Huge swarms of these insects converged on the lions, causing skin lesions, exsanguinating them, and leaving emaciated lions barely able to hunt. According to Henry Fosbrooke, conservator of the Ngorongoro Crater during the outbreak, the population plummeted from a high of one hundred lions to a low of ten individuals, as tormented lions perished or ran from the crater in horror.

Packer was fascinated with this event because it was a real-time population bottleneck in an isolated population. He reckoned that he could reconstruct the entire recovery of the lions with a pedigree based on pride association and individual identification. Craig and Ann had developed an elaborate lion identification scheme based on the patterns of whisker spots that allowed them to recognize and track individual lions. Every field biologist who had tracked lions in the crater had taken close-up photographs of each lion they saw, providing a photo record of the living crater lions since 1962. To back these up, the Packers put out classified ads in African tourist magazines for amateur lion photos from the thousands of people who had visited the crater between 1962 and 1978, just before he and Ann began their oversight. He figured that nearly every tourist who spotted a lion in the Ngorongoro Crater would try to take a picture! And he was right.

It took years of painstaking analysis and sorting through thousands of lion photographs for Craig to trace the lions' recovery since the *Stomoxys* plague. But eventually he determined that all the modern crater lions were descended from fifteen founders, eight survivors of the *Stomoxys* plague and seven immigrant males who wandered into the crater in the few years following the epidemic. It was difficult to know the fathers of cubs precisely, but the pride mothers could sometimes, but not always, be identified by cub-rearing association. For the first time, a pedigree of close inbreeding following a defined population bottleneck had been documented precisely. We were very anxious to examine whether a molecular genetic and reproductive analysis of the Serengeti and Ngorongoro Crater lions would affirm the bottleneck's effects and reveal more details about the history of these lions.

The National Geographic Society bankrolled our team's expedition to Tanzania in 1987. Eight of us boarded the plane for East Africa along with twelve huge trunks of veterinary drugs, blood- and skin-sample-processing supplies, bush clothing, and very precious nutrients unavailable on the Serengeti plain, like peanut butter, apricots, and Triscuits. We landed in Nairobi, where we purchased more supplies, organized vehicles, gathered multiple spare tires, and were joined once more by wildlife photographer Karl Ammann. Karl's curiosity and unfettered eagerness to help in all areas more than made up for his lack of scientific training. He had proved invaluable in our previous safaris looking for cheetahs in the Masai Mara and the Serengeti. Also, he could spot cheetahs and lions one hundred times better than anyone else we knew.

Our convoy set off for Seronera, a tiny village in the center of the Serengeti Park that researchers like Craig and Ann used as a home base. Craig had previously installed radio collars on a dozen lions. Since lions sleep all day, they were usually easy to find, dart, collect biological specimens from, and resuscitate. The vets would always wait for the lions to recover enough to walk, run, and defend themselves before leaving. After a week of very long days in the Serengeti we had specimens from twenty-seven lions, enough for an evaluation. We headed next for the Ngorongoro Crater.

Things began rather well at the crater study site. During a planning trip about six months earlier, my pretty and always smiling wife, Diane, had persuaded an upscale tourist outfitter, Abercrombie and Kent, to set up a tent camp for us on the crater floor for a tiny fraction of their standard cost. Diane convinced them that they wanted to be part of this landmark study. They took it on.

Up at dawn each morning, Craig and the two wildlife vets, Mitch Bush from Washington's National Zoo and Don Janssen of San Diego Zoo, went out in search of lions to dart. The first day went very well; three lions were darted, sampled, and recovered. We relaxed with hot showers and a gin at our fancy safari camp, tired from a long, hot day wrestling lions for their bodily fluids. Then it all unraveled.

A park ranger appeared at our camp early the next morning and informed us that the park conservator, Mr. Joseph Kayera, had ordered the lion collection stopped immediately. The ranger gave no explanation, just simply delivered the order. I climbed aboard our rented Land Rover, took a Motorola hand radio, and began the treacherous two-mile rocky "up road" from the crater floor to the crater ledge where the park office was located. Kayera was not in and I was told that his busy schedule might prevent him from seeing me today, or at all, for that matter. I insisted that I would wait cheerfully.

Day became dusk, then evening, and still no conservator, so I made my way to the Ngorongoro Park Lodge and booked a room. The lodge is a beautiful luxury hotel with a view of the entire crater floor. At the bar sipping Tusker ale that night, a group of tour-bus drivers revealed the source of our troubles. A princess from the Netherlands had visited the day before and became extremely distressed upon seeing a bunch of characters "molesting" the lions in the crater. She was not sure what was happening, but the lions certainly looked dead to her. The princess, of course, shared her horror with the conservator. My work was cut out for me.

The next morning I radioed the camp with the grim news and headed for the park office. Kayera appeared at around ten A.M., but he was too busy to see me. I waited patiently until four P.M., when he invited me into his office. He welcomed me and proceeded to describe in some detail his disdain for the way that arrogant American researcher Dr. Craig Packer was carrying out his program. He never sought permission for collars, did not report his findings, harassed the lions, frightened the tourists, and, just like all the other expatriates, was insensitive to local regulations and culture. Kayera was furious. He did not really know or care much who I was, but by association with Packer, he assumed I was no good for this place either. The princess's complaint was simply the straw that broke the camel's back. The project was canceled! Go home! Now! Please!

I was tired, intimidated, and stunned. But I decided to tell him what he wanted to hear. Yes, Dr. Packer was not perfect, certainly arrogant and insensitive. Of course he should communicate more with his Tanzanian hosts. But I implored him to understand that this project was bigger than all of that. The veterinary assessment alone would benefit the animals and inform managers. The anesthesia and sampling were extremely safe. So far every lion had recovered properly; none had died. I promised to redress his fears, side with him on all fights against Craig. (I assumed Craig would forgive me for the sake of the project.) I promised to explain calmly to any visiting tourist how a biomedical evaluation was an invaluable management and research tool of great benefit to the lions.

By nine P.M. he had softened; the project could continue, but only for one day at a time and with the requirement that I report back to him daily on all aspects. I thanked him for his insight and wisdom, went to the crater's edge, and radioed the team to go ahead. For the next several days Kayera and I became friendly as I spent many hours detailing progress and problems. The Dutch princess left Tanzania and our team collected samples from sixteen crater lions, enough for a very close analysis.

Once back in the States, Wildt, Bush, and I were quick to examine the specimens. We first looked at the Serengeti lions, which displayed a great deal of allozyme genetic diversity, easily as much as outbred house cats or other wild cats like ocelots or leopards. Measures of DNA variation at the lions' MHC, the gene complex we had measured by skin grafting in cheetahs, were also highly variable. Instead of performing skin grafts (hardly an option with wild lions), we measured the lions' MHC with a technique called "restriction fragment length polymorphism," RFLP, which tracks DNA sequence differences in the genes that encode MHC proteins.

As we suspected, the crater lions presented a very different story. They had 50% less overall molecular genetic diversity than the prides of the Serengeti. In other words, their 1962 population bottleneck had cost the population half of its endemic genetic variation. Dave Wildt found that sperm counts of crater males were only 60% of that of Serengeti lions. The larger outbred Serengeti lion population had about 25% abnormal sperm per ejaculate, a sign of rather healthy sperm, while crater lions had twice that number. The hormone profiles run on the lions' serum samples showed a dramatic difference as well. Serum testosterone levels of Serengeti lions were three times higher than the testosterone levels in crater males. Testosterone is a critical hormone produced in the testes that mediates sperm development. Low testosterone concentration in crater lions was the likely reason for the elevated frequency of malformed sperm. Inbreeding after the 1962 bottleneck was having the effect on the crater lions we had feared and Craig's elaborate pedigree reconstruction had predicted.

The grim situation of the crater lions, however, would seem modest alongside the faraway lion population we would study next. The lions of the Gir Forest Sanctuary in the Gujarat state of western India comprised a relict population of about three hundred animals that were the sole survivors of the Asiatic subspecies of lion, *Panthera leo persica*. Formerly occupying a vast range from Turkey and the Arabian Peninsula on the east to western India/Pakistan on the west, the Asian subspecies was extirpated by agricultural development and rampant colonial big-game hunters. Census records from 1880 to 1920 show multiple periods, even multiple generations, when the population dropped to fewer than twenty individuals. Once lion hunting was outlawed in the Indian state of Junagadh in the 1920s, the population increased gradually to its present size. Today it occupies a sanctuary area, fourteen hundred square kilometers in size, in the Gujarat peninsula of western India.

Asiatic lions look different in several respects from African lions. They are a bit smaller, and most have a marked skin fold running along the length of their underbelly. Males have a very shortened mane, and about half of Asian lion skulls, including all of today's Gir lions, have a bony ridge in the cheekbone that crosses an opening for nerves to the eye called the "infraorbital foramen." In all other felid species and in African lions, the foramen is a single opening with no bridge.

These physical characteristics were originally thought to be adaptations or at least modifications associated with the long time that Asiatic lions were isolated from African lions. Our molecular genetic estimates would suggest that African and Asian lions have been isolated from each other for at least fifty thousand years. We now believe, however, that these physical traits in Asian lions are manifestations of extremely severe inbreeding in their very recent past. The evidence for our conclusions was encrypted in their genes.

Paul Joslin, the deputy director of the Brookfield Zoo in Chicago, had spent three years tracking lions in the Gir Forest Sanctuary, but he was more worried about the troubled population of captive Asiatic lions. In 1981, he had helped establish a Species Survival Plan (SSP) for Asian lions under the auspices of the Zoo Associations of America, Australia, and Europe. Captive Asiatic lions in Western zoos had been bred and managed as a population backup for the tiny wild population since the early 1980s. By 1989, 205 Asiatic lions were being bred in thirty-eight different zoos. Paul explained that he was concerned because many of the captive offspring did not show the diagnostic belly fold, the small mane, or the infraorbital foramen bridge. Also, the entire captive population was descended from only five founders, a tiny number that might constitute an inbreeding threat. Furthermore, the records of the precise parentage of those five original founders were suspect, particularly since two founders came from an Indian zoo that was rumored to interbreed African lions with Asian lions. The Indian zoo officials denied the charge.

Paul offered to organize blood sample collections from the captive population and from the Sakkarbaug Zoo, just outside the Gir Forest in Junagadh, India. The Sakkarbaug Zoo lions were assuredly authentic, since they had all come right out of the adjacent Gir Forest.

The Gir Forest Sanctuary borders on heavy human settlements, inevitably leading to lion attacks on local citizens. Ten of about one hundred attacks in the 1980s were fatal. Wildlife agents captured all the man-eaters and placed them in the Sakkarbaug "breeding program"—a sort of maximum security penitentiary for the killer lions. Joslin wanted to find out for sure if his captive SSP lion population was "pure" or "hybrid." I wanted to look at the Gir lions' genes to compare them to their African cousins.

Bush, Wildt, Joslin, and I set out for the Gir Forest in hopes of collecting authentic Gir lions both from the Sakkarbaug Zoo and from the Gir Sanctuary itself. The Sakkarbaug Zoo was overflowing with twenty-eight lions, largely because human attacks were on the rise. We collected blood, semen, tissue, and serum from all the Sakkarbaug animals carefully and quickly.

The free-ranging lions in the Gir Forest Sanctuary were more of a challenge. The teak forest habitat is dense, extremely arid, dusty, and malaria-infested. For the first time in our lives, our team was tracking lions on foot; the forest was too thick for a Land Rover. And this was a place with a dozen lion attacks on people each year. The park rangers carried only narrow wooden spears for protection—no guns, blow darts, or pepper spray—so when we would come across several lions lying in a creek bed, we were very cautious. Looking back now, it seems inconceivable that we were able to anesthetize and sample the lions, but somehow we did succeed with six males.

Sadly, a freak accident during our study in the Gir Forest led to a young lion's death. The park rangers had lured a small group of lions to a clearing using a slaughtered gazelle for bait. Four lions were chewing on the carcass when we arrived. Bush darted a young male who jumped as the blow dart injected into his back leg. A nearby female, also eating, reacted to his jump with a swat and then bolted off into the forest. The darted male followed in hot pursuit. We moved slowly around the other lions and then took off on foot after the two lions. They had run over a mile on the dense dusty game trail to find perhaps the lowest point and the only body of water within miles, a tiny basin two meters wide and a foot deep. The tranquilized male succumbed to the drug precisely at the water hole, collapsed, dropped his head under the water, and drowned. He had died seconds before we arrived.

The moment was tragic and devastating to witness, even as we tried to persuade ourselves that the small risk we always take when we anesthetize wild animals served a greater good. Bush performed a complete autopsy. We filed a detailed written report of the accident. Although our Indian hosts were understanding, supportive, and even forgiving, we did not have the heart to continue the collections any further. We returned to the United States with samples from six wild males and twenty-eight captive Gir lions from the Sakkarbaug Zoo.

The genetic profile we drew from these samples was rather disturbing. Asiatic lions from the zoo or the forest had virtually zero measurable genetic diversity. Not one of the fifty allozymes was variable, no MHC-RFLP variation was apparent, and the mitochondrial DNA sequences measured by the RFLP technique were all identical. This was much worse by far than the Ngorongoro Crater lions. A dramatic affirmation came from the population pattern of minisatellite markers, the same highly variable repetitive DNA sequences that we had used to estimate the date of the cheetah's population bottleneck. Minisatellite DNA patterns are resolved by gel electrophoresis and they resemble supermarket bar codes, with every individual so different as to be unique in a population. The extensive variation of human minisatellite loci is the DNA fingerprint used in murder and rape cases. Serengeti and even Ngorongoro crater lions showed rich diversity in DNA fingerprint patterns, but the Gir lions were all identical. It was as if they were all clones or identical twins. This was the most genetically uniform population we had ever observed. The Gir lions had even less variation and were more severely inbred than cheetahs.

The historic inbreeding that homogenized the Asian lions' genes led to some rather dramatic physiological consequences. David Wildt's reproductive analysis showed that Gir lions had a tenfold diminution in sperm count compared to the Serengeti lions. The incidence of malformed sperm was 66% in Gir lions on average, compared to 50% in crater lions and 25% in Serengeti lions. Gir lions had five times fewer motile sperm per ejaculate than Serengeti lions and a tenfold reduction in serum testosterone levels. The testosterone depletion in these lions would cause not only the increased malformed sperm in every semen sample, but also the dramatically underdeveloped mane. The Gir lion males had become "feminized" by their history of inbreeding.

The reproductive distress of the male Gir lions seemed to have translated into breeding troubles as well. Sakkarbaug lion pairs frequently failed to conceive or produced stillborn cubs. Cub mortality was much higher in Sakkarbaug compared to captive African lions in other zoos. Even the rare normal-appearing sperm from Gir lions' ejaculate were defective in fertilization tests attempted in Dave Wildt's laboratory.

Add to these reproductive defects the Asian lions' distinctive traits—the reduced mane, belly fold, and infraorbital foramen bridge—and we see a recipe for inbreeding depression. The genetic evidence for historic inbreeding was overwhelming and the cost of these events was unmistakable.

Remember the captive SSP population of Asian lions in America that Paul Joslin worried about? When we examined the genetic structure of that population we found it to be quite different from the Gir and Sakkarbaug lions. The Asiatic lion SSP population retained quite a bit of intrinsic genetic variation, in stark contrast to the genetically monotonous Gir lions. Also, the genetically variable "alleles" of the SSP animals were familiar ones, because we had seen them previously in African lions. When we examined the SSP lion pedigree carefully and tracked the inheritance of these variable African alleles, we became convinced that two of the five original founders of the SSP Asiatic lions were actually from Africa. The entire SSP lion pedigree was mongrel. Thirty-eight zoos were engaged in amassing a "pure Asiatic" lion population that in truth derived from matings between African and Asian lion forebears. Oops!

Paul and other zoo managers of the Asiatic lion SSP were not happy. Their very successful breeding program, with far better fecundity and productivity than the "pure" Gir lions from Sakkarbaug, was a genetic admixture of two continental subspecies. As the bearer of bad news, I tried to emphasize what I believed was a positive spin on the revelation. Of course the SSP lions did well: They had inadvertently ameliorated all of the woes of inbreeding present in "pure" Asiatic lions. The pure Gir lions were intrinsically flawed, reproductively impaired, and weakened by generations of inbreeding. The SSP population offered a living testament to the real benefits of maximizing outbreeding. I urged the zoo community to perpetuate the SSP lions as an experimental population to allow further, more extensive scientific study and as a reminder of the benefits of good genetic management.

Nobody listened to me. My optimism notwithstanding, within a year of our disclosure, all the SSP lion "hybrids" were fitted with birth control implants! Nobody wanted to conserve a hybrid subspecies. This was my first, but not my last, close encounter with the politics of species and subspecies hybridization.

Naturally, we wondered how the Gir Forest lions had become so genetically depressed. They clearly had descended from a very severe, long-lasting bottleneck. But when was it? Was it the result of documented overhunting by British big-game hunters a hundred years ago, or was there more to it? How long did it last? We had a way to answer these questions using a very simple but elegant concept called the "molecular clock hypothesis."

In the early 1960s evolutionary biologist Emile Zuckerkandl teamed up with two-time Nobel Prize winner Linus Pauling and put the following idea forward. They reasoned that when the population of a species splits apart, perhaps by migrating across a huge river or mountain range, the descendants of the split-up populations would gradually change over time by acquisition of new mutations in their DNA sequences. As time passed, more and more mutations randomly dispersed across DNA stretches would accumulate. The longer the time elapsed, the greater the gene sequence divergence.

If we could measure the amount of sequence difference for the same gene between two populations (these are called "homologous" genes since they descend from a common ancestral gene), that amount of difference would be proportionate to time elapsed since they split from their ancestors. This means that hidden in the gene sequences of all living species are stopwatches that reveal the time of separation from related species. By comparing the same homologous DNA region (say, the hemoglobin gene) between lions and tigers, we see a measurable difference, but the same gene difference between tigers and bears would be twenty times greater because the common ancestor for cats and bears would be much further back than the ancestor of tigers and lions.

This proportionality of DNA sequence divergence and elapsed time forms the basis for the field of molecular evolution. DNA sequence divergence can assess the ancestral relationship between living species. Applying molecular data to related species is totally revolutionizing our understanding of the historic connections among living species. In coming chapters I will describe how the molecular clock has breathed new life and increased precision into taxonomy, the science of species classification, with examples from cougars, pandas, and orangutans. Today's evolutionary biologists have not only fossils and morphological variation to inform them, but also the clocklike DNA molecules as heralds of ancient species divergences.

A particular challenge for those of us who try to read the molecular clock is picking the correct gene sequence. Some genes accumulate mutations very slowly, one or two in ten million years. Other genes evolve more rapidly, sometimes up to a hundred times faster. Slowly evolving genes are useful for very old divergences such as the rise of mammals 70–100 million years ago, but useless for more recent population splits on the order of thousands of years ago. For recent events, rapidly evolving genes must be used.

I was guessing that the Gir lions' population bottleneck was around a hundred years ago due to widespread big game hunting, but I wanted to be sure. To examine this event a very committed and con-servation-minded graduate student, Carlos Driscoll, had selected a group of DNA segments that we knew evolved very, very fast: genomic microsatellites. Carlos was convinced these markers would hold the answer to the Gir lions' mysterious past.

Microsatellites are short, stutterlike sequences found in chromo-somes where a pair, a trio, or a quartet of nucleotide letters are repeated in tandem at least a dozen times. A microsatellite "locus" is a repeat stretch found at a specific chromosome site. All the mam-mals examined so far possess between 100,000 and 200,000 microsatellite "loci" distributed in a nearly random manner across the entire genome. Because the cell's DNA photocopying machinery makes mistakes frequently when it encounters a microsatellite locus (the genetic equivalent to a typographical error), the mutation rate for microsatellites is a thousand times higher than it is in nonrepeated DNA or gene coding regions. Because of their propensity to collect spelling errors, nearly every microsatellite locus has accumulated many alleles, between five and thirty in most populations.

But this was not the case in the Gir lions, where seventy-one of the eighty-eight microsatellite loci we tested were invariant—one allele, the same in every lion. When we looked closer, however, there was an interesting paradox. If the severe bottleneck that hammered the Gir lions' ancestors occurred around 1900, then there should be no vari-able microsatellite loci at all, particularly since all the other genetic measures were completely flat. A mere century, fewer than twenty lion generations, is not enough time for mutation to regrow more than one or two new microsatellite alleles, even with their relatively high mutation rate. Yet Gir lions showed variation in seventeen microsatellite loci of the eighty-eight Carlos tested, each retaining two, three, or four alleles. How could that be?

When we considered the known mutation rate for microsatellites, we calculated that it would take around three thousand years for a severely bottlenecked population (one where every microsatellite locus was reduced to a single invariant allele) to reconstitute as much new variation as we were seeing in modern Gir lions. The bottleneck that compromised the Gir lions' genetic variation dated back not just one century but three millennia!

A second measure also relating to microsatellite locus variation helped confirm this estimate. When a microsatellite locus diversifies by new mutation from a single to many alleles, the size range between the smallest and largest alleles increases with time. So if we measure the maximum breadth or size difference between alleles at each microsatellite locus and then take the average size span across all the microsatellite loci, that average breadth is proportional to the time elapsed since the bottleneck. The average allele size span was measured across all seventeen variable microsatellite loci of the lions and compared to the same loci in another species that also survived a bottleneck, the African cheetahs. For Gir lions, the average microsatellite size difference was about 18% of that for cheetahs. Since the cheetahs' genetic bottleneck occurred 12,000 years ago, the Gir lion bottleneck computed as 18% × 12,000, or 2,100 years ago. The Gir lions' genetic reduction began a few thousand years ago, pre-dating by 2,000 years the nineteenth-century big game hunters.

A look back at the geological history of the Gir peninsula helped us understand the new estimate. Around twenty-five hundred years ago the Gir peninsula was actually an island surrounded by rising water. The ancestral founders of the Gir lion population were isolated from the larger mainland lion population, and due to small numbers they suffered inbreeding over several generations. Hunting and habitat occupation by human developments had effectively extirpated the larger mainland lion populations over this period, leaving only the inbred Gir lions to occupy the area once the peninsular waters receded.

Not only did our genetic tools further cement our conclusions about the perilous history faced by endangered species, but they were also able to weigh in on nattering questions about the lions' social organization, puzzles that had become burning issues for behavioral ecologists since Schaller's original Serengeti lion study. Because Craig and Ann Packer knew all their lions personally (or at least by their whisker spots), they were anxious to see if once and for all they could test their theories about lion-mating and cub-rearing behaviors. While we were busy collecting blood samples from the Packers' study prides, they patiently anticipated the parentage results. Who were the actual parents of each cub?

This was not a simple question. Lion mothers rear their cubs so communally that the mother of any single cub is obfuscated. Mothers mate with multiple resident males repeatedly, obscuring the father's identity as well. Craig believed that exact knowledge of parentage would either support or refute prevailing hypotheses about lion behavior, particularly those grounded in the established evolutionary concept that transmitting one's genes is the only driving force in nature. Oxford zoologist William Hamilton had coined the term *kin selection* to describe a component of natural selection in which relatives behave in a manner that promotes their genes to survive through matings of their closest kin. Put simply, if evolution is all about transmitting one's genes successfully, would there not be an advantage to helping your brothers and sisters transmit theirs? Lion pride organization offered a perfect chance to investigate this theory, since for lions, everything sexual is a family event.

Dennis Gilbert, an extremely talented graduate student, accompanied us to the Serengeti on our first lion expedition and took on the task to solve uncertain maternities and paternities in the lion prides. Back in our NCI laboratory, Dennis had isolated feline DNA-fingerprint minisatellite segments from house cat DNA. He used these to demonstrate the extreme genetic uniformity of the Gir lions, but he also applied the lions' minisatellite loci to assess specific mother and father identification for some eighty lion cubs born in Craig and Ann's Serengeti lion prides.

Dennis and the Packers used these assignments to nail down several conclusions that would make the Serengeti lion study a paradigm for wildlife behavioral ecology. First, Dennis's paternities proved that resident males in a pride fathered all cubs and no outside males had snuck in to breed with females, as is seen commonly in chimps, birds, and humans. Second, as expected, all the females of a pride were very close relatives (sisters, cousins, aunts, mothers, and daughters), meaning females never allow nonrelative additions. Third, the males in a pride coalition were never closely related to the females, demonstrating a natural inclination of the lion prides to avoid breeding with their kin. So far, no real surprises, only hard genetic affirmation of behavioral suspicions.

But what about the male coalitions? Nomadic male coalitions ranged in size from single males to large groups of five or six lions. Years of observation had shown that the single most important determinant of who wins a takeover contest is the relative size of the battling coalitions. Larger coalitions nearly always prevail. The question was, are all members of a male coalition brothers or close relatives like the females? Or are males willing to join up with nonrelatives to increase their likelihood of success in takeover attempts? Further, once a pride residency is won, which of the coalition males actually fathers the offspring?

Dennis's parentage assessment answered all these questions. About half of the male coalitions were made up of only brothers and the other half were mixtures of brothers and unrelated males. But there was an important pattern: All large coalitions of males were exclusively brothers and only small coalitions (of two or three lions) included unrelated males.

What about reproductive success? It turns out that although all the lion males will copulate with estrus lionesses, in nearly all litters, just two of the males sire all the offspring. This result seemed to explain two phenomena: first, the nonchalance of males in waiting their chance to mate, either because their rival is a close relative, as in larger male coalitions, or because in small coalitions they have a very good chance of siring cubs. Second, single- or doubleton males may join up with nonrelatives to ensure takeover success because being in a small group does not greatly diminish their chance of siring cubs. But large consorts never join up with outsiders. That way even if a male in the larger coalition fails to father any cubs, his brother, who carries 50% of his genes, will. So Hamilton's kin selection theory is borne out in lion prides. Lion brothers assist their brothers in gene proliferation. The group cooperation is an effective strategy for transmitting each male lion's genes.

One cannot help but marvel at regal lions, not only for their unabashed majesty, but also for the remarkable lessons they have taught us. The Serengeti lions serve as a prime example of the adaptive benefits of maximal outbreeding, while the Ngorongoro Crater and Gir Forest lions show the enormous genetic cost of close inbreeding. The price of inbreeding has been described elsewhere in the scientific literature but never quite as thoroughly as in these lion populations. The lions have offered by happenstance a precisely controlled natural experiment, one we were fortunate to document.

The graphic picture of communal survival strategies documented in lions also has broad implications. George Schaller makes the point in his Serengeti lion monograph that possibly more can be learned about the acquisition of human social systems by studying lions than from studying evolutionarily closer but asocial solitary monkey species. After all, humans were subject to the same natural experiments in which only the adaptive—that is, successful—strategies prevailed. Fully understanding these subtle field interactions and their perturbations could offer a fresh perspective on human behavior and sociality.

There remain unanswered questions about lion society that I know will be addressed by bigger and better studies in the future. One particular puzzle that deserves attention is why lions are the only cat species to cooperate so intensively in their social systems. All other cats live solitary, isolated existences. The cheetahs' hermit lifestyle in the African savannah has surely protected them from spreading the inevitable epidemics that would exploit their immune disease gene uniformity. Parasites and fatal viruses rely on intimate physical contact to spread through a population. So how are lions protected from such a risk? Have they somehow evolved an immune system shield that fills the defensive void of genetic homogeneity, or could a new disease lead to their extinction any day?

There are no easy answers to these questions; however, we have made more sense of them by integrating components of adaptive behavior, immunity, and reproduction. Our studies of the lion and the cheetah have taught us a bit about these great cats and raised some thoughts about our own species. These beautiful animals brought together specialists from very different science disciplines, enabling us to mine secrets we might never have encountered alone.

Four

A Run for Its Life—The Florida Panther

When the last individual of a race of living things breathes no more,
another heaven and another earth must pass before such a one can be
again.
—WILLIAM BEBE

NONE OF US WERE MUSICIANS, BUT ON THAT HUMID, SULTRY
evening in October 1992 our discourse grew to as intense a crescendo
as any diva had ever delivered. The Florida panther conservation
workshop had convened in a most elegant spot, the White Oak Plan-
tation, an eight-thousand-acre millionaires' paradise on the Georgia
border near the Osceola National Forest of rural Nassau County,
Florida. White Oak is an elegant vacation retreat, a thoroughbred
breeding center, and a hospice for endangered species established by
philanthropist and paper magnate Howard Gilman. The setting is
reminiscent of California's Hearst Castle in its grandeur. White Oak
director John Lukas, a committed conservation advocate, had invited
scientists, ecologists, government managers, and conservation NGOs
(nongovernment organizations) there for a single purpose: to break
the Florida panther recovery plans out of a logjam of academic
squabbles and bureaucratic inaction before one of the most severely
endangered animals in the United States went extinct.

Thirty concerned citizens, each committed to panther conservation, listened, postured, criticized, struggled, and compromised to

evaluate available data, imperfect as they were. We needed to make major decisions on how to rescue a population of less than forty Florida panthers clinging to survival in the inhospitable alligator- and mosquito-infested Big Cypress Swamp, just north of the Everglades. Somehow, the group reached a consensus, if not a unanimous one: immediately supplement the fragile populations with animals from a neighboring subspecies, the Texas cougar, in a last-ditch effort to reverse the almost certain extinction of the Florida panther. As the resolution came through, I could not help but notice the emotion emitting from the face of my friend, Dr. Melody Roelke, who years earlier had organized the skin grafting of cheetahs in Oregon. By now, Roelke had served a seven-year tenure as field veterinarian dedicated to the Florida panther. With a stream of tears running down both cheeks, she whispered softly, "At last they are finally listening to me." Indeed they were.

The saga of the Florida panther was disturbingly familiar: a tiny population in limited habitat, hemmed in by human development, suffering from inbreeding and dying by road kills, illegal hunting, depleted prey base, and other catastrophes. However, this population enjoyed exceptionally close scrutiny. Maybe too close.

The Florida panther was first listed as endangered by the U.S. Department of Interior in 1967, and the Endangered Species Act of 1973 mandated its protection. In 1982 Governor Bob Graham designated the panther Florida's state animal after it was chosen by school children over alligators and manatees. Two federal government agencies, the National Park Service and the U.S. Fish and Wildlife Service, had joined with the Florida Department of Natural Resources and the Florida Game and Freshwater Fish Commission to monitor the panther population, to assess the immediate threats and hopefully reverse its imminent extinction. The cumulative research, field observation, and oversight provided the conservation community with perhaps its most vivid glimpse ever of the process of population collapse. And all of the incremental data, decisions, debates, and actions occurred in a media feeding frenzy that awarded anyone with an opinion, however unfounded, headline status. Scores of books and newspaper and magazine articles have been penned to weigh in on the recovery effort, not to mention thousands of pages of internal government reports over a twenty-year interval. Nearly buried in all this cacophony lies the thread of this lonely, mysterious creature's remarkable story.

New World "lions" were first reported in the annals of Christopher Columbus's fourth voyage along the seacoasts of Honduras and Nicaragua in 1502. During his three-thousand-mile expedition in 1539, Hernando de Soto described "lions" living in such abundance that resident Native Americans in Florida posted guards on their burial grounds after dark to keep the scavengers away.

The Florida panther is a common name for a subspecies of puma, which until the end of the nineteenth century was widespread in the southeastern United States. The puma species was named *Felis concolor* or "cat of one color" by Carolus Linnaeus, founder of the Latin-name species classification scheme, in 1758. The puma was described as a handsome, ecologically versatile American big cat that can be found in deserts, in savannahs, in tropical rain forests, and on alpine steppes. Today, pumas range from the Canadian Yukon, south through the Rocky Mountains of the western United States, into Central and South America to the southernmost tip of Patagonia. Pumas, also called mountain lions, cougars, catamounts, and scores of other names, can thrive at sea level or in the Rockies and Andes at elevations above thirteen thousand feet. They are secretive and elusive predators that evoke conflicting emotions of terror and admiration from people who encounter them in the wild.

The Florida panther was first described scientifically in 1890 by ornithologist Charles Cory from the Field Museum in Chicago, who christened it a puma subspecies: *Felis concolor floridana*. Another nineteenth-century biologist, Outram Bangs, elevated the panther to full species status, honoring Cory by dubbing it *Felis coryi* (the name *Felis floridana* had already been taken by bobcats). This convention stood until 1946 when Stanley Young and Edward Goldman reviewed puma records and reports and named thirty-two different subspecies of puma, including the Florida panther (*Felis concolor coryi*) and the Texas cougar (*Felis concolor stanleyi*), named for Stanley Young. Today the Florida panther is considered a puma subspecies and is the only surviving population of pumas east of the Mississippi.

The American South during the seventeenth and eighteenth centuries saw the growth of the great plantations, numerous towns, and sprawling tobacco farms. Wolves and panthers were hunted to near extinction for sport, to protect livestock, and simply because they were perceived as dangerous varmints. Peninsular Florida lagged behind in agrarian development largely because its acidic sandy soils and dense swampland were less than ideal for cultivation. This temporary slowing of human expansion allowed the panther to persist in Florida long after it had vanished from other southern states.

However, once the Gulf Coast of Florida began to develop during the first half of the twentieth century, panther numbers plummeted. There were so few in the 1950s and 1960s that many believed them extinct. But no one was quite sure.

Anecdotes of panther sightings, shootings, plaster of Paris casts of panther tracks, and eyewitness encounters crept into Florida newspapers. In March 1969 a local sheriff near the central Florida town of Inverness shot a Florida panther. Then a motorist hit a puma south of Moore Haven. The World Wildlife Fund bankrolled U.S. Fish and Wildlife Service zoologist Ron Nowak and a lanky West Texas puma hunter, Roy McBride, to search for any remaining Florida panthers. McBride, who was to become the lead houndsman for the Florida panther recovery team, trained smallish Walker foxhunting hounds to sniff after the trail of big cats and chase them up into trees. In February 1973, Nowak, McBride, and McBride's dogs treed a large female puma in Glades County, along Fish Eating Creek, west of Lake Okeechobee. That episode triggered a serious movement to find and protect a relict population of panthers living in Florida.

In his gripping book *Twilight of the Panther,* Ken Alvarez, longtime Florida panther advocate and Florida Department of Natural Resources biologist, chronicles how the official government-sponsored conservation efforts began in March 1976 in a Unitarian Church in Orlando. There, a handful of biologists, game wardens, and conservationists reviewed the evidence for panther survivors and agreed to staff a Florida Panther Recovery Team under the auspices of the U.S. Fish and Wildlife Service. Florida Game and Fish biologist Robert C. (Chris) Belden was nominated as chair, and he began a comprehensive quest to assess the status of the surviving population. For several years, Belden led a team in search of panthers, drawing on the hunting expertise of Roy McBride. They treed and darted their first cat in February 1981. The big cat was fitted with a radiotelemetry collar for remote radio-tracking. Several other panthers were found and within a few years, the crew estimated a population of around twenty individuals. The age ratio was troubling; it revealed nearly all of the panthers were old, near geriatric. Was this the last generation?

Belden's team collected life history data: weight, age, appearance, and medical samples for parasites. They noted that the animals had three physical traits that were unusual for pumas captured west of the Mississippi but were nearly always found in the wild Florida panthers. The first was a mid-dorsal whorl or cowlick on the back of the neck. The second was a 90° angle turn in the terminal vertebrae of the tail, a crook or kink. The third was a patch of white flecks on the back of the neck between the shoulders. The original "type specimen" of Florida panther collected by Outram Bangs in the nineteenth century had none of these traits. His Florida panther description emphasized cranial morphology, specifically the presence of a distinctive wide flat frontal region of the skull with broad, highly arched nasals, long limbs, and a rich rusty color. Belden found Bangs's identifying characteristics to be vague and inconsistent; the more objective kinks, cowlicks, and white flecks rapidly grew to be the standard for Florida panther identification. Every panther had the tail crook and nearly all had a cowlick, while less than 5% of western cougars had these traits. The white flecks later turned out to be tick-induced, that is, nongenetic, and were subsequently dropped.

Two years after the first capture, the unthinkable occurred. Belden's team darted a panther, which suddenly died of heart failure after a heated chase while it was being lowered from a tree. The public reaction was loud, accusational, and broadcast like a bullhorn in Florida newspapers: "Leave panthers alone! Stop harassing the panthers!" Chris Belden's years of field experience, dedication, and acumen were irrelevant in the court of public opinion. He was reassigned to a desk in Gainesville. For safety, a highly trained wildlife veterinarian was added to the team. That veterinarian, fresh from her involvement with the African cheetah's medical-genetic secrets, was Melody Roelke.

The field team was headed by biologist John Robotsky, aided by houndsman Roy McBride, and medically monitored by Roelke. Robotsky had invented an inflated crash bag that circled the tree of the hunted panthers' refuge to catch falling panthers. Roelke set up an elaborate field lab to perform physical exams and to collect blood, feces, throat swabs, and sperm. Within two years, Robotsky was replaced by field ecologist Dave Maehr, who served as team leader for eight years. Maehr's 1997 book *The Florida Panther* spins a rich narrative of the perils and perspective of the lead ecologist charged with conserving the immensely popular, high-profile cat.

By 1986, the team had a dozen collared adults and the population was estimated at thirty to fifty individuals. Roelke invited David Wildt to monitor the reproductive physiology of the panthers and asked me to look at the genetic structure of the tiny population at risk for both extinction and inbreeding. Over the next several years, a chilling portrait of the struggling population, the great-great-grandchildren of a proud, dense, wide-ranging eastern panther, came into focus.

With every capture, Roelke shipped us blood and skin tissue specimens by Federal Express. We used the serum to look for antibodies to pathogenic cat viruses. The skin snips were digested with enzymes and planted on plastic tissue culture bottles to grow living panther cell lines. Allozymes and DNA markers for population diversity were assessed from red blood cells and white blood cells.

The molecular genetic results were as we had feared. The Florida panthers were all genetically compromised. The tiny population had a fivefold reduction of allozyme variation compared to several populations from the American West, the lowest of any North American puma subspecies and nearly as low as what we had seen in cheetahs. DNA from mitochondria showed the same level of genetic uniformity, as did minisatellite DNA fingerprint assessment, the forensic

bar code variation used in the Gir lions. Florida panthers were showing signs of close inbreeding and historic shedding of intrinsic genetic
diversity, a red flag for an endangered species. When combined with
Belden's traits, tail kinks and the cowlick, the Florida panther had all
the earmarks of a population damaged by a population bottleneck.

Dave Wildt's reproductive analysis was telling. Sperm of Florida
panthers had a ratio of 6% normal to 94% developmentally malformed sperm. This was the worse skew we had ever seen, worse than
cheetahs, worse than Gir lions, and very much worse than puma subspecies from South America. Over 40% of Florida panther spermatozoa showed a defective acrosome, a critical structure in the collar of a
sperm cell. Acrosome defects completely incapacitate sperm for normal fertilization. Further, the average count of total sperm per ejaculate in Florida panthers was between eighteen and thirty-eight times
lower than in other puma subspecies and thirty times lower than what
we had seen in free-ranging cheetahs. A puzzling observation that
I shall return to in a moment was a surprisingly high incidence of
malformed sperm, around 70–80%, in the pumas or mountain lions
that lived in the Rocky Mountains. This level of abnormal sperm
approached that seen in cheetahs and Gir lions, but was not as bad as
the 94% abnormality seen in Florida panthers.

The veterinary surveillance provided by Melody Roelke and her
assistants revealed additional problems that correlated with and I
believe were also consequences of inbreeding. As more and more kittens were captured and monitored, the incidence of cryptorchidism
was seen to rise from 0% in 1970–74 to 90% of the births in the early
1990s. In cryptorchid males, one testicle fails to descend to the scro-
tum, leading to reduced sperm production, or in the case of bilateral
cryptorchidism (where both testicles fail to descend), complete sterility. The condition is hereditary and likely reflects the expression of a
causative recessive mutation. The specific mutation in Florida panthers has not yet been identified but a genetic basis for the condition
has been confirmed in several domestic species, namely dog, swine,
sheep, and domestic cats. Cryptorchidism was found in 56% of all
Florida panther males examined, but was never observed in medical

exams of over forty pumas captured in Texas, Colorado, British Columbia, and Chile.

By 1994, some fifty-five Florida panthers that died during the years of the study were examined by autopsy. In 1992, a cardiac abnormality termed "patent foramen ovale," whereby the heart's atrial valve fails to close properly, appeared and caused the death of five panthers. Roelke diagnosed one panther with the defect by noting an extremely offbeat heart murmur during a routine physical. She would describe the stethoscope sound as similar to an unbalanced washing machine chugging across a room. That animal was the product of a mating between a father and his daughter. In all, 80% of Florida panthers showed heart murmurs compared to less than 4% of other puma subspecies.

As if this were not enough, the infectious disease burden on Florida panthers is enormous, much higher than in other puma subspecies. Over fifteen microbial agents have high prevalence, including rabies virus and *Pseudomonas aeruogenosa,* microbes that are uncharacteristic of healthy animals and are seen to occur in hosts with a disarmed immune system (e.g., in people on immune-suppressing drugs or suffering from AIDS). At least eight recorded panther mortalities were attributed to microbial agents. As we suspected for cheetahs, the inbred history and the isolated lifestyle appear to strike a survival balance for the Florida panther. Inbreeding increases susceptibility to fatal microbes, while isolation limits the spread of these agents.

The Florida panthers were displaying the most extreme physiological problems I had seen in any endangered species. The suspicion that inbreeding was the cause gained credence through a rather unusual and perhaps serendipitous occurrence. In 1986 a small family group of panthers was captured in the Everglades eighty miles south of the Big Cypress Swamp population. None of the Everglades panthers had tail crooks and only one out of seven had a cowlick. When we examined mitochondrial DNA (using the RFLP method), every

Everglades cat had a genetic type quite distinctive from the type found in the Florida panthers of the Big Cypress Swamp. A comparison of the Everglades' mitochondrial DNA genotypes to other pumas showed they resembled—in fact were identical to—a puma subspecies from Costa Rica. So what happened? How did a Costa Rican genetic lineage get to South Florida?

We were puzzled, but we knew of a captive population of Florida panthers raised in a roadside zoo called the Everglades Wonder Gardens in Bonita Springs, Florida. This group was established by the zoo director Les Piper from wild Florida panthers he captured in Hendry County in the early 1940s. Piper never admitted it, but there were persistent rumors that in the 1950s he had introduced some non-Florida pumas from somewhere "south of the border" to improve his breeding. Our molecular genetic analysis of seven of Piper's captive panthers revealed that some had the authentic Florida panther mitochondrial DNA while others had the Everglades' Costa Rican type. Both Piper and Everglades cats had a distinctive South/Central American allozyme marker, *APRT-B,* that was completely absent from the authentic Florida panthers living in Big Cypress or from any western U.S. cougars. The genetic link was made, suggesting that somehow the captive Piper animals had made their way into the Everglades. But how did they get there? Roelke needed to get to the bottom of this puzzle.

She rummaged through the old files of the Everglades National Park Ranger's Office and discovered the answer. She found a 1965 letter from the park administration to Les Piper thanking him for his cooperation in the release of seven of his captive stock of Florida panthers into the Everglades. A long-forgotten government-sanctioned panther reintroduction project had left a genetic trace in their descendants. This was our smoking gun. The trail had in the meantime disappeared and no remnant of the animals surfaced until the field team captured their great-grandchildren in the Everglades twenty years later.

What Roelke and I uncovered was a serendipitous experiment where a genetically impoverished "authentic" Florida panther had mated to another subspecies from Costa Rica and had successfully returned to the wild. And what about the physiological correlates to inbreeding that were so obvious in the Big Cypress panthers? They were mostly cured in the Everglades panthers. The kinks and cowlicks were gone; there was no cryptorchidism or cardiac abnormalities. Their sperm was still rather malformed, but we believe this represents the genetic influence of the authentic Big Cypress Swamp panthers.

Unfortunately, the Everglades panther family was short-lived. By the summer of 1991, the last Everglades female perished, probably due to toxic levels of mercury found in her postmortem tissues. Two remaining males migrated north to the Big Cypress Swamp in search of a mate, while one male wandered alone in the Everglades. The natural experiment provided hope but ultimately the Everglades panthers lost their battle. The remainder of the war for survival was to be waged in the Big Cypress Swamp, a larger and better habitat for the surviving Florida panthers dealt a genetic curse of inbreeding.

Probably because conservation management issues are rather new, we sometimes draw conclusions in critical cases without the benefit of a complete set of facts. It is not surprising that most observers of the plight of the Florida panther were convinced that a few centuries of human depredation and development was responsible for the cats' precarious situation. Will the Florida panther go the way of the Everglades cats, the dodo, or the Caspian tiger? This question can better be answered if we go back a few steps to examine more closely the cause of the panther's predicament.

Consider what we do know about the natural history of pumas. Some five million years ago in North America, the ancestors of modern American pumas split to form three different felid species: the puma, the cheetah, and the jaguarundi, a small cat now found in South and Central America. These cats and a few others like the saber-toothed cats and American lions (real lions of the genus *Panthera*) represented the continent's dominant carnivorous predators, nourished by a rich fauna of elk, deer, and buffalo. Around two million years ago the two American continents were joined for the first time. This allowed the migration of a multitude of species into South America, particularly carnivores like the puma. Before the continents joined there were no placental carnivores in South America, only marsupial (related to kangaroos) carnivores left over from the ninety-five-million-year-old contact between South America and Australia in the ancient supercontinent Gondwanaland. Pumas thrived over this wide range, as evidenced by their frequent appearance in the fossil record of 400,000 to 500,000 years ago.

However, around 18,000 years ago, the last ice age rendered all of Canada and the northern tier of the United States a several-mile-thick ice sheet comparable to modern Antarctica. As the ice melted over thousands of years the barren landscapes gradually transformed into the rich forested terrain and grassland prairies of North America. When the big thaw was nearing an end a puzzling event occurred. Between 9,000 and 12,000 years ago, a wave of mass extinctions abruptly eliminated forty species of mammals from North America. These Pleistocene extinctions, named for the geological epoch of the period, involved predominantly large animals, mammoths and mastodons, and five successful carnivores: the dire wolf, massive short-faced bears, American lions, saber-toothed tigers, and cheetahs.

Similar large extinction waves also occurred in South America, Australia, and less extensively in Eurasia and Africa. Nearly all the North American extinctions involved large mammals, although a number of large flesh-eating birds, eagles, vultures, condors, and teratorns were also eliminated, never to appear again, save in our imagination.

The cause of these great extinctions, the most extensive in the seventy-million-year fossil record for mammals, is a mystery. The two prevailing guesses, climatic environmental pressure or the destruction caused by human immigration in these regions, are at a stalemate. Nearly all regions of extinction were briefly inhabited by early humans, which is why some feel they played roles. I was not there at the time, so I can only speculate, but surely a global catastrophe of some sort triggered the cataclysm. It probably was not a galactic meteorite, as we believe finished the dinosaurs sixty-three million years ago, because the timing and duration of the large mammal extinction was different in different continents. My suspicion is that a devastating epidemic swept through the large herbivores, reducing their numbers, leaving large carnivores and birds of prey starving. It could have been caused by a pathogenic agent that jumped from live-stock that were beginning to become domesticated by the people liv-ing at that time.

Fatal epidemics are fairly common; witness the catastrophic rinderpest outbreak that diminished East African buffalo and wilde-beest by 95% in the late nineteenth century. There are countless other examples. Human predation and environmental crises certainly may have played a part in the late Pleistocene extinctions, but the global plague scenario makes more and more sense as microbe hunters learn about these scourges from modern species.

So how did the puma survive the pressures that wiped out other North American species? The answer seems to be that North Ameri-can pumas actually did not. The genetic evidence for an alternate story was uncovered by a determined and extremely thorough gradu-ate student, Melanie Culver, who began her Ph.D. research hoping to resolve taxonomic disputes over puma subspecies classification.

From the start, Culver and I were skeptical that the thirty-two sub-species named by Stanley Young and Ed Goldman had a firm genetic basis, so she attempted to examine the subspecies issue using the most powerful genetic tools available, mitochondrial DNA sequence analysis and a collection of nuclear microsatellite repeats that appear throughout the chromosomes of mammals. Melanie used the same microsatellite markers we had used to date the bottlenecks of chee-tahs and lions to search for unique marker alleles that could confirm the distinctiveness among subspecies. With help from Melody Roelke's network of puma field biologists, plus the contacts, acumen, and persuasive charm of a fresh young postdoc, Warren Johnson, Culver would spend several years collecting puma blood and tissues from each of the named puma subspecies. When she could not get blood samples from puma hunters and mountain lion biologists, she settled for a skin snip from a museum specimen or hunter's trophy. Over a decade Culver, Roelke, and Johnson collected some three hundred puma tissues including a number of museum skins from nineteenth-century Florida panthers and the extinct eastern cougars, a subspecies that once roamed through New England and the mid-Atlantic states.

Although Culver's passion was species conservation, her hidden strength was genetic biotechnology. She developed new methods for extracting DNA from tanned hides and for assessing DNA sequences for mitochondrial and microsatellite segments. The DNA sequence from three of the thirty-seven genes in the mitochondrion (*16s rRNA, ATP8*, and *ND5* of 315 pumas, a total of 891 base pairs) gave her a clear picture of puma population structure. Melanie and Carlos Driscoll, her lab partner and an energetic new breed of conservation geneticist, assessed over 120,000 microsatellite genotypes of pumas and other cats for comparison. Their data set was massive, the evolutionary analysis daunting. But the results were definitive, relevant, and provocative. Culver presented her findings at the annual meeting of the American Genetic Association in the summer of 1999.

Culver found that the thirty-two separate geographic subspecies were impossible to affirm or distinguish as separate populations even with the most advanced molecular genetic tools we had. Most named subspecies were part of a large population with continuous gene flow between them. The best she could do was to detect genetic subdivision among six geographic groups: 1) southern South America, including Chile and southwestern Argentina; 2) eastern South America, including Brazil, south of the Amazon, and Paraguay; 3) northern South America, including Venezuela, Ecuador, Bolivia, Guyana, and Colombia; 4) central South America, including northeast Argentina; 5) Central America, including Costa Rica, Panama, and Nicaragua; and 6) North America, including the United States, Mexico, and Canada.

On close inspection, puma populations in Central and South America were rather typical of other large carnivores in genetic structure. They showed a lot of genetic diversity in mitochondrial DNA and in their microsatellites. A distinctive pattern of mitochondrial DNA genotypes discriminated between the five geographic locales below Mexico. The microsatellite results pointed to the same conclusion, five geographic populations below Mexico and one group up north.

Each southern puma group showed abundant variation, up to ten times what we saw in the Florida panthers. We recommended that the five southern groups be recognized scientifically and legally as subspecies validated by distinctive DNA-based genotypes.

The puma populations in Mexico, the United States, and Canada were more puzzling. In this vast continental region that encompassed the range of sixteen traditional subspecies named by Young and Goldman, all the pumas were nearly identical in genetic terms. One hundred animals shared a single mitochondrial genotype and four cougars from Vancouver Island had a second genotype that differed only by a single nucleotide letter from the more common type. The microsatellite markers told the same story even more graphically. North American pumas were markedly similar, showing twenty- to fiftyfold less genetic diversity than South American puma populations. The results indicated that all of the North American pumas had acutely homogenized genetic diversity. This could only mean that the entire North American population descended from a severe population bottleneck.

Melanie Culver and Carlos Driscoll estimated the time of the North American puma's population bottleneck by measuring the microsatellite allele size range, the same approach Driscoll had used to date the Gir lions' bottleneck. That size range expands with time by the slow accumulation of new mutation-generated alleles after a homogenizing population bottleneck. The average allele size range seen across eighty-five microsatellites in North American pumas was virtually identical to the same measurement we had seen for African cheetahs. This meant that North American pumas had experienced their bottleneck at about the same time as the African cheetahs. Since the cheetah's bottleneck date was determined to be 10,000 to 12,000 years ago, then North American pumas were all descended from a few individuals that lived at about this same time. The microsatellite allele size range in South American pumas was much greater than the range for North American pumas, setting the age of their intrinsic diversity at over 40,000 years ago.

The implications of these calculations were provocative. All North American pumas alive today—in the Rockies, Canada, Mexico, and

Florida—all trace back to a handful of founders that lived around 12,000 years ago, exactly the time of the great Pleistocene extinctions. Pumas had migrated from South America into North America at that time following the mass extinction event that had eliminated so many large mammals and predatory birds. Yet, fossil remains for pumas in North America extend way before this cataclysm, raising the question, What happened to the pumas that occupied North America before the late Pleistocene extinctions? The answer: Whatever extirpated American lions, saber-toothed tigers, mastodons, giant ground sloths, and American cheetahs also wiped out the entire continent of resident pumas. This was one big extinction event.

After the loss of large mammals, large cats, and pumas from North America, a tiny band of pumas, perhaps a family group or two, moved north to establish home ranges in the regions vacated by the previous, less fortunate inhabitants. The immigrant pumas would establish and defend large territories, then disperse adolescents northward. The number of founders remained effectively low by virtue of behavioral competition; resident pumas living in the Panama isthmus and northward blocked any further immigration of southern-born pumas. With time the north-spreading descendant pumas settled in Mexico, the United States, and Canada, creating a new greater North American population, but one with a dramatically reduced sampling of their ancestral South American genetic diversity.

The Florida panther's genetic depletion therefore actually began 12,000 years ago to be exacerbated by a second severe bottleneck in the twentieth century. Culver's analysis of 100-year-old museum specimens of the Florida panther revealed five times more overall genetic variation than today's Big Cypress Swamp panther. The nineteenth-century Florida panther showed comparable genetic diversity to western cougars, with both populations reflecting the late Pleistocene genetic reduction. The consequences of the Pleistocene population bottleneck are still apparent and likely are the cause for the high frequency of malformed sperm (70–80%) that David Wildt had seen in the pumas of the western states. These elevated sperm abnormalities are much higher than were observed in the outbred South American pumas, which show less than 40% abnormal sperm.

The double whammy that afflicted the Florida panther was worrisome to say the least. It was in this context that John Lukas invited scientists, field biologists, administrators, and conservation activists to White Oak in October of 1992.

By anybody's perspective, and there were plenty, the prognosis for the fragile population of panthers was bleak. Even though the contiguous habitat of over two million hectares spanning South Florida was as large as any wildlife reserve in the eastern United States, it was still too small for a healthy puma population. Chance mortality from road kills, common diseases, and hurricanes could easily finish the population. The genetic data were even worse news: Not one but two near brushes with extinction had shed the Florida panther's protective genomic diversity and it showed. Panthers carried not only disfiguring benchmarks of inbreeding—tail crooks and cowlicks—but also life-threatening cardiac abnormalities, cryptorchidism, extreme sperm defects, and a remarkably high incidence of microbial disease. Ulysses Seal and Robert Lacy, population biologists with specialties in conservation policy, used computer models based on the present breeding structure of the wild population to forecast the near certain statistical likelihood of the panther's demise through accidental events in fewer than fifty years. Left alone, the Florida panther would be remembered as a textbook exercise on how to go extinct while your abundant and vociferous advocates argue about the process.

The workshop participants at White Oak adopted a joint resolution: to act and act soon. The executive summary of the proceedings recommended that "reinstitution of historic gene flow between the Florida panther and adjoining [puma] subspecies, i.e., genetic augmentation, is needed to reverse the [population] and inbreeding effects," and the "action needs to be taken quickly." Three possible scenarios were given top priority: direct translocation of pumas from another region to Florida, artificial insemination of Florida females, and artificial insemination of non-Florida pumas with sperm collected from wild Florida panthers.

There were some dissenting opinions on the matter. One came from Florida conservationists with no faith in government-assisted conservation efforts. They were absolutely convinced that any manipulation of panthers for any reason was harassment and detrimental. These well-meaning and emotionally charged advocates had attended every Florida panther meeting or workshop since the first panther capture mortality a decade earlier.

Another determined opposing voice came from the leader of the field team, ecologist Dave Maehr. After three years in the field Maehr had developed his own firm impression that the population's single greatest threat was limited habitat. He was unconvinced that panthers suffered from the consequences of inbreeding and once suggested that the ill-fated Everglades cats may have succumbed to "outbreeding depression," a long-since-dismissed concept raised to justify antimiscegenation laws in the Old South.

Maehr's insight was comprehensive and genuine, but his unflappable adherence to ecological concerns set up severe conflicts in his mind. Maehr became quite agitated over the risks of panther anesthesia versus the benefits of increased time of sedation for blood and semen collection and biomedical assessment. He asserted his opinions from field observations over executive dictums by government administrators, managers, and bureaucrats, and his disdain for holding a caged panther for assisted reproduction or captive propagation. He spurned computer models, instead calling for increased field data. These tensions made Maehr the chief opponent of "genetic augmentation," but he was one of only a few who were persuaded to dissent from the consensus recommendations.

Nearly three years of government ambivalence and inaction followed the White Oak resolution. Finally, a Florida congressional delegation coerced the U.S. Secretary of Interior to implement the genetic augmentation plan. In the spring of 1995, Roy McBride translocated eight large nonpregnant healthy cougar females from West Texas and released them in different spots in the Big Cypress Swamp. The number was set to achieve a 20% genetic augmentation of the current population of thirty to fifty breeding adults. Translocation of animals was chosen over other options because artificial reproduction technologies were not yet developed as a reliable technique for pumas. Texas cougars were selected because as recently as 1900 the Texas subspecies range overlapped that of Florida panthers, so nineteenth-century gene flow was simply being reconstituted. In fact, the name of the relocation action was changed to "genetic restoration," at least in part for public relations.

Over the next several years, the introduced Texas cougar females mated with Florida panther males. By mid-2001, seventeen first-generation and thirty-three second- and third-generation intercross panthers were born in the Florida habitat. The intercross offspring are apparently strong and healthy, more robust than their "authentic" Florida panther fathers. Tracker Roy McBride described them as supercats with strong muscles and sleek energetic sprints, making them tougher for his dogs to run down and tree. Once treed, the muscular hybrids would leap over the dogs and the field team to escape, a spectacle rarely seen with their authentic but visibly weaker Florida panther parents and grandparents.

The density of panthers in the Big Cypress habitat has nearly doubled, fulfilling the hopes of the White Oak gathering. The offspring do not have tail crooks or cowlicks and the males examined so far show diminished sperm abnormalities, comparable to western cougars. The restoration action cannot be proclaimed as an outright success yet, but it certainly is moving toward that. The many of us watching the large natural experiment are keeping our fingers crossed that maybe one day the panther's adventure will be considered a conservation success story.

The Florida panther saga will be remembered for years to come but not only in conservation circles. Determining that North American pumas descended from a late Pleistocene bottleneck is a vivid reminder that the genomes of modern species have built-in chronometers, ones we can use to interpret their natural history. We are getting better at reading these signals, just as paleontologists have learned to reassemble long-extinct species from fossil fragments. Slowly, but in exciting small increments, an adventure like the Florida panther saga teaches us more and more how to read the mysterious code encrypted in genes.

Humankind may derive an unexpected benefit in addition to the continued presence of a beautiful creature. The cardiac arterial septal defect seen in Florida panthers is among the most common congenital heart defects seen in children. It is repairable by surgery, but suspicions of its genetic basis gain credence from the Florida cats. As geneticists work to unravel the human genome and the several million single nucleotide variants that distinguish each and every person, we are nearing the day when connecting specific gene variants to heart disease, congenital aberrations, and hypertension—diseases with clear genetic influence—will be routine. The Florida panther reminds us that specific gene identification offers a feasible path for new diagnosis and treatment of heritable diseases.

We learned similar lessons for human reproduction. The clear impact of recessive genes on sperm aberrations or testicles retained in the abdomen in cats can help identify the genetic basis of these defects in ourselves. The incidence of human male infertility is enormous. Geneticists are beginning to identify the genes that, when mutated, impair human reproduction in metabolic pathways that parallel the wild cats. When these new discoveries translate to treatments, they will owe no small debt to the beautiful and fragile panthers of Florida.

In the end, the Florida panther gave us as close a look at an endangered species on the brink of extinction as had ever been documented. The whole exercise also tested the ability of conservationists, scientists, and government agencies to work in concert. A rhetorical struggle ensued between two diametrically opposed ideologies, one insisting that we leave the poor animals be, let nature take its course, and the other side pressuring for immediate action. Both camps claimed passionately to represent the panther's best and only hope for survival.

In my narrative of the Florida panther, I have glossed over some rather prickly management issues so that the scientific story could shine. But the science of threatened species does not operate in a vacuum; conservation decisions are all connected to people, society, government, law, and public policy. In the next chapter, I will explore the critical decision-making processes and behind-the-scenes political frenzy that dogged our scientific efforts to protect the endangered panther subspecies.

Five

Bureaucratic Mischief

IN 1983, PERHAPS AS A REACTION TO THE FUROR IN FLORIDA newspapers over the death of the Florida panther captured by Chris Belden's team, Florida's governor, Bob Graham, appointed a five-person Florida Panther Technical Advisory Council of government and academic scientists to oversee panther conservation actions. One of its members, John Eisenberg, a distinguished professor at the University of Florida, was widely renowned as an expert on all aspects of mammal biology. Eisenberg had strong opinions on most things, and his opinion on Florida panther identification demonstrated remarkable prescience. Eisenberg had persuaded his colleagues on the Technical Advisory Council that strict adherence to the tail crook, white flecks, and cowlicks as formal criteria for Florida panther recognition was politically perilous.

In 1986, Eisenberg and Chris Belden exchanged their perspectives on Belden's favorite diagnostic characters, just a few months prior to the first capture of the newly discovered Everglades group. Eisenberg asked Belden, "What are you going to do if the panthers captured in the Everglades do not have cowlicks or crooks in their tails? Will they qualify as Florida panthers or not?" Those who overheard the exchange thought Eisenberg's concerns were unfounded. But, indeed, the Everglades pumas did not have tail crooks or cowlicks. Further, they had molecular genetic markers more reminiscent of pumas in Costa Rica than in Florida. Two lineages of pumas were

roaming in Florida and the Everglades version enjoyed a little accidental genetic augmentation. Was this good news or bad news?

Melody Roelke was too excited by the genetic sleuthing to worry
about political implications. Roelke's only concerns were the health
of the panthers and the pursuit of scientific truth. But as the data
became clearer and more certain, some of Roelke's colleagues challenged her enthusiasm for the new results, not because of their
keener scientific acumen, but because of the ramifications for panther conservation. They urged Roelke and me to consider suppressing
the results—simply to not disclose them publicly. The reason:
Hybrids between endangered species and other taxa would be ineligible for protection or funding under the U.S. Endangered Species
Act. The entire Florida panther recovery effort would hang in the
balance of the disturbing new genetic findings.

Eisenberg and his committee had urged the U.S. Department of
Interior and the Florida Game Commission to revise the definition
for the Florida panther as a cushion against such a possibility. The
two agencies reacted as typical bureaucracies and did nothing. Now
Melody and I held a smoking gun that could abruptly derail support
for one of the few endangered species to have received continuous
government funding and protection. Ken Alvarez estimated the cost
of two decades of panther conservation expenditure at something
over thirty million dollars.

Taxonomy, also called "systematics," is the science-based hierarchical
classification of the world's species. The area had traditionally been
an obscure academic discipline dominated by erudite and professional dons who would memorize and interpret thousands of Latin
species names. Advances seldom made the newspapers and caustic
disputes lingered in the dusty scientific literature for generations.
That academic innocence would be lost forever when precise taxonomic recognition of species and subspecies came to be the basis for
protection under the Endangered Species Act.

Sometimes referred to as the plant and animal's "bill of rights," the

Endangered Species Act was signed into law by Richard Nixon in 1973. Over one thousand U.S. species and some five hundred foreign species are listed as endangered by the U.S. Fish and Wildlife Service, which today has a backlog of thirty-five hundred species awaiting elevation to protected status. The act is forty-five pages long and specifies that no one is to take, harass, harm, pursue, hunt, shoot, wound, kill, trap, capture, or collect a listed species. The penalties are stiff, including jail terms and up to $25,000 in fines per day. The law is considered one of the most effective environmental protection documents ever enacted anywhere.

Over its thirty-year history the Endangered Species Act has weathered assaults of every possible variety. As recently as 1996, bills were introduced in both houses of the U.S. Congress to scrap it or at least revise it. In 1978, the Supreme Court affirmed the redirection of a $100 million dam in deference to the tiny endangered snail darter in Tennessee. President George Bush Sr. dismissed the act as "a sword aimed at jobs, families, and communities of the entire region of the U.S." His secretary of the interior, Manuel Lujan, described the Endangered Species Act as "a gross impediment to economic progress that needed to be . . . repealed." In 1995, Supreme Court Justice Antonin Scalia, in a minority opinion regarding protection of habitat for the spotted owl in Sweet Home, Oregon, argued that "it is not necessary to protect breeding sites, because impaired breeding does not injure living creatures." Gale Norton, George W. Bush's secretary of interior, who oversees the act, once argued unsuccessfully to the Supreme Court that the law is an "unconstitutional infringement of private property rights." The act has violently polarized wildlife environmentalists and land developers since its inception.

The strange saga of Chief James Billie, chairman/chief of Florida's Seminole Indian tribe, illustrates the intricate complexities that evolved relating the Endangered Species Act to the Florida panther conservation. In December of 1983, the forty-year-old folk singer and Vietnam veteran was spotlighting deer with a friend in a pickup truck on the Seminole reservation in the Big Cypress Reserve when the green eyes of a panther reflected the searchlight's beam. Billie shot the cat with a high-powered pistol, skinned it, made a meal of the animal, and hung the skin on a cypress pole near his home. Alerted by an informer, wildlife wardens from the Florida Game Commission arrived the next day and confiscated the pelt and some panther remains in the cooking pot. They arrested James Billie. He was indicted under U.S. federal charges and Florida State statutes for killing and possessing an endangered species, a Florida panther.

The Florida media portrayed the ensuing court case as an epic battle between two hyperliberal causes: the right to protect endangered species versus the innate right of Native Americans to hunt freely or to carry on their lifestyle irrespective of the federal laws. Billie's lawyer argued that Seminoles were not subject to wildlife regulations on their reservation, where hunting is permitted year-round. Indeed, he proclaimed, there would not be an endangered Florida panther had it not been for exploitation of Florida by the white man.

There was no question that Billie had shot the cat. The skin was on his pole and panther bones were found in his cooking pot. However, when the case came to trial in U.S. District Court in LaBelle, Florida, in October 1987, the defense came armed with a more potent argument based on disputed taxonomy. They asked, If panels of experts could not agree on what a Florida panther was, how could a jury convict a private citizen for shooting one?

Melody Roelke and Laurie Wilkins testified for the prosecution. Wilkins, a professional taxonomist from the Florida Natural History Museum, had recently examined 648 pumas from twenty-nine puma subspecies and analyzed the very high incidence of tail crook, cowlick, white flecks, and cranial characters on living and historic museum specimens of Florida panthers. She testified that Billie's cat had a tail crook, a cowlick, and arched nasals similar to the authentic Big Cypress Florida panthers.

In his summation, the defense attorney argued deftly. The question of identifying Florida panthers was disputed even among the "experts." Of the three Florida Game Commission markers used to identify Florida panthers, one was nongenetic, the tick-infested white flecks, and the other two, cowlick and tail crooks, had never been mentioned by Cory, Bangs, and Goldman, the pioneering zoologists who originally described the Florida panther. And word was out that the Everglades cats lacked these diagnostic characters. How could a common loyal citizen identify the protected panther? Who were the Florida panthers anyway?

The federal judge instructed the jury that any conclusion that "waivered or vacillated" constituted a "reasonable doubt" and grounds for acquittal. The federal case was dropped after a hung jury terminated deliberations. Billie was acquitted shortly thereafter in the state court proceeding. The next year, our laboratory received tissue from Billie's trophy and determined it to be an authentic Big Cypress Florida panther based on mitochondrial and nuclear gene analysis. Today both James Billie and O.J. Simpson live as free men in Florida.

In December 1990, our genetic results detailing the distinct lineage of Everglades versus Big Cypress panthers were scheduled to appear in the now-defunct scientific journal *National Geographic Research*. Neither my coauthors, the field team, nor I were comfortable with the possibility that the data might be used by the enemies of conservation to subvert invaluable federal protection for the panther. We were aware that considerable confusion existed over definition and recognition of endangered species. Adding to the confusion were the new, very powerful but difficult to understand molecular genetic technologies that could reveal, with amazing precision, long-forgotten historic events like the Everglades cat's background. Was there a way to extract the Florida panther and the new hybrid lineage data collected on behalf of conservation from this legislative catch-22? Our position was untenable: publish the results and unleash a legal loophole for adversaries of endangered species or suppress important scientific data. Neither option was acceptable as far as I was concerned. I had to find a way out.

I began with a crash course on the history of the Endangered Species Act and its applications and interpretations, all published in the Federal Register. The Act had been designed to protect species whose number and habitat had become diminished sufficiently to threaten their survival. The law as written protects three kinds of taxa: species, subspecies, and specific vertebrate populations.

Between 1977 and 1983 the Solicitor's Office of the Department of Interior, which serves as legal counsel for the U.S. Fish and Wildlife Service, issued a series of tortuous internal memoranda to deal with hybridization of endangered species. The dusky seaside sparrow, southern Selkirk Mountain caribou, and sea otter were the objects of these opinions. The Solicitor's Office ruled in each case that hybrids between endangered species, between endangered subspecies, or even between listed populations would not be protected under the Act's provisions. The memos concluded that the protection of hybrids would not serve to recover listed taxa and would more likely jeopardize their survival even further. This opinion, which came to be known unofficially as the "Hybrid Policy," had the force of judicial precedent and would influence management and conservation decisions for years to come.

The dusky seaside sparrow, a melanistic subspecies of sparrow that inhabited central and eastern Florida, was listed as endangered in 1966. By 1980 the population had plummeted to six birds. Five were brought into captivity to be crossbred with a lighter-colored Scott's seaside sparrow from the Gulf Coast of Florida in a heroic effort to save the dusky. The Solicitor's Office vetoed the experiment, citing previous Hybrid Policy memoranda. The dusky went extinct.

A few years later molecular genetic studies showed retrospectively that the subspecies that was crossed to save the dusky was probably a poor choice as they belonged to populations that had been isolated for a very long time in subspecies terms, over 250,000 years. Ironically, and luckily, the neighboring Atlantic Coast subspecies which survives well today was shown to be genetically very closely related. The dusky was not so unique or endangered after all. So in a way, the dusky never really went extinct.

Other fallout from the Hybrid Policy was not so innocuous. In 1990, Robert Wayne, the student who had measured the lopsided cheetah skulls, now a UCLA genetics professor, completed an elegant genetic study that showed that wolves and coyotes formed hybrids in nature in a number of locales when the two species overlapped their ranges. Almost immediately after hearing this, the legal staffs of three western state farm bureaus, Wyoming, Montana, and Idaho, formally petitioned the U.S. Department of Interior to remove the grey wolf from the endangered species list, because it formed hybrids with the common non-endangered coyote. Their argument quoted directly from the Solicitor's Hybrid Policy opinions. The U.S. Fish and Wildlife Service denied the petition, arguing that the wolf-coyote hybridization was very rare. Actually it is not so rare, but the case illustrates how the Hybrid Policy painted the government into its own small corner, one that required conjured-up arguments to defend endangered species protection.

The scientific basis for species and taxon recognition by the U.S. Fish and Wildlife Service is a paradigm in taxonomy called the biological species concept, or BSC. The BSC was formulated in 1940 by Ernst Mayr, a Harvard ornithologist and arguably the most influential evolutionary biologist since Charles Darwin. Mayr's insight and his clear articulation of wide ranges of zoology, biology, ecology, and systematics put him in the vanguard of biological thinking and policy for his entire career.

Mayr's BSC defines species as groups of actually or potentially interbreeding populations that are reproductively isolated from other such groups. The operative words are *reproductively isolated*. There have been dozens of other species definitions proposed that defined species based on evolutionary potential, their past history, or selected diagnostic characters observed in the species. But Mayr and his proponents preferred the BSC because it seemed objective and definitive. The species themselves evolve intrinsic barriers that naturally prevent crossbreeding in the wild. By contrast, definitions based on similarity of visible characters depend on which characters are used. Because the BSC had long been widely known and accepted, the U.S. Fish and Wildlife Service adopted the concept for its enforcement of endangered species legislation. This convention would become the foundation for the Hybrid Policy.

Over the past sixty years a rather impressive assembly of respectable taxonomists and evolutionary biologists have tried to unseat the BSC for a wide variety of reasons. Most of them failed, probably because Ernst Mayr is alive, adroit, and articulate at ninety-six years young as I write these words, and most critics are no match for him.

In early 1990, just as our panther dilemma became apparent, another giant in evolutionary genetics at Harvard, Richard Lewontin, asked me to present a lecture to his graduate students. Lewontin was a pioneer in the estimation of population variation, being the first to estimate overall genetic diversity of *Drosophila* populations from allozyme variants in 1966. I dropped Professor Mayr a short note before my visit and asked if he might have time to meet briefly to discuss the predicament that the BSC-based Hybrid Policy had created for species like the Florida panther and the grey wolf. He agreed and I was thrilled.

I spent a week reading up on various species concepts, including several of Mayr's books, in anticipation of our meeting. When the day arrived I knew I would have only an hour of his time. I explained the dichotomy of Florida panther lineages, the wolf-coyote intercrosses, and the Hybrid Policy. I asked him to collaborate with me in writing an editorial piece to resolve in simple terms that natural events embodied in the BSC should support, not defeat, conservation goals. With his cooperation I was hoping we could influence or even dismantle the Hybrid Policy, not by an emotional appeal but by a reasoned scientific perspective on the natural implications and interpretation of the BSC on species and subspecies activities in the real world outside the Washington beltway.

Professor Mayr was very pleased to help and our meeting extended for an extra hour as we exchanged thoughts, perceptions, prejudices, and ideas. We would compose several drafts in the coming days, and within a few weeks, we had a draft essay that we submitted to *Science*. It had an intentionally prickly title: "Bureaucratic Mischief: Recognizing Endangered Species and Subspecies." To get the right attention, we needed a headline with a punch.

Science reviewed and accepted the article; Mayr and I braced for public and government reaction. The paper first outlined the problem posed by enforcement of the Hybrid Policy—that it threatened the protection of several species and subspecies. With the BSC originator as its coauthor, the paper powerfully called upon Mayr's 1988 corollary that "since hybridization may occur occasionally between individuals belonging to different good species, it is important to stress that the term reproductive isolation refers to the integrity of populations, even though an occasional individual may go astray." We pointed out that in nature limited "hybrid zones" occur in hundreds of cases when closely related species come into contact. Yet, such hybrid zones do not disintegrate the genetic *integrity* of either parent species, nor should they confuse the recognition of species that endure them.

We also noted that species and subspecies are very different things. Although species are reproductively isolated, subspecies are not. Adjacent subspecies quite commonly connect and merge their gene pools. This is what we see in nature and even in human populations where different ethnic groups—African, Asian, Caucasian, the human equivalent of subspecies—intermarry freely.

Our essay offered a new extended definition for subspecies as "geographically defined aggregates of local populations which differ taxonomically from other species subdivisions." To help government regulators recognize subspecies, we suggested explicit guidelines. Members of a subspecies share a unique geographical range or habitat, a group of recognizable genetically controlled characteristics, morphological or molecular, and unique natural history as compared to other subspecies. Since they are not a distinct species, they are reproductively compatible and will periodically interbreed with adjacent subspecies. We reasoned further that all subspecies have the potential to acquire suitable adaptations to their specific ecological situation and the longer they are separated the more cumulative adaptation we might expect. All subspecies have the potential to one day evolve into new species as described in 1866 by Charles Darwin; unfortunately it is never possible to know which subspecies will fulfill this potential.

On its surface the Hybrid Policy makes sense for fully developed species that produce sterile hybrids when they breed together, like when lions breed with tigers or horses with donkeys. No one would like to encourage such unnatural hybrid matings. However, for subspecies and populations, the Hybrid Policy should be scrapped. Subspecies, we reasoned, have at least four possible natural fates. They can gradually change to a new subspecies, they can go extinct, they can evolve into a new distinct species, or they can meet another subspecies and exchange genes, an intercross. All are common natural outcomes in the wild—so endangered subspecies like the Florida panther or dusky seaside sparrow should not be penalized for what happens naturally every day.

I crossed my fingers hoping that the pointedness of our opinion, the preeminence of my coauthor, and the timeliness of the topic would break through the inertia at the U.S. Fish and Wildlife Service. A few days before the article appeared I received calls from several reporters covering the scrap. One correspondent, William Stevens of the *New York Times,* told me that he had just called the spokesman of U.S. Fish and Wildlife for a comment. A few weeks earlier, I had sent the agency officials a courtesy preprint of the editorial so they would be prepared. The spokesperson emphatically told Stevens, "There is no Hybrid Policy. The whole matter is being rethought here. . . . The service is now in the process of developing a policy as to what is and what is not a hybrid."

I was amazed. They had revoked the Hybrid Policy in advance of our article. It was the best news I could have imagined. Victory within the system! We had adjusted, however slightly, a vulnerable loophole in the interpretation of the Endangered Species Act. The Hybrid Policy was no more, yanked a few weeks before our editorial appeared, effectively removing the legal basis for suspending protection for the Florida panther.

This was good news for scores of other endangered species. In Washington State the northern spotted owl had been crossbreeding with a neighboring subspecies. Hybrids between blue whales and fin whales were discovered based on new molecular genetic data. Even the red wolf, a flagship endangered species for U.S. Fish and Wildlife Service that, like the Florida panther, had received continuous protection and funding for over a decade, was shown to derive completely from hybridization between coyotes and an extinct wolf subspecies. Indeed, the time was right to rethink the problem.

Eighteen months later, the White Oak Florida panther workshop would recommend its Florida panther restoration plan. The decision to restore new blood into the sickly population was made possible by the Hybrid Policy reversal. With an intact Hybrid Policy, reintroduction of a different subspecies from Texas not only would have been controversial, it would have been illegal. Now that the policy was under review, even the government officials had become less uncomfortable with the restoration plan.

Remember that 150 years ago the range of the Florida panther abutted that of Texas cougars. The scheme was merely restoring natural gene flow between adjacent races or subspecies that had become interrupted by human settlements. This argument, paired with the urgency of the situation, gave the cautious decision makers the ammunition they needed to defend a bold and sure-to-be-controversial move.

It took the U.S. Fish and Wildlife Service five years to create a new hybrid policy. On February 7, 1996, the new policy finally appeared in the Federal Register, under the title "Endangered and Threatened Wildlife and Plants: Proposed Policy and Proposed Rule on the Treatment of Intercrosses and Intercross Progeny (the Issue of Hybridization)." The new policy was very straightforward, but cleverly crafted. It simply recognized as endangered "hybrid individuals that more closely resemble a parent belonging to a listed species than they resemble individuals intermediate between their listed and unlisted parents." Protection of endangered taxa would "indicate the inclusion of intercross and intercross progeny individuals within the original listing action for the parent entity. The policy is meant to aid in recovery of listed species by protecting and conserving intercross progeny." The new rules required that criteria for recognition be developed through an approved recovery plan including a genetic management plan. They acknowledged the damage to endangered species posed by the withdrawn Hybrid Policy memoranda from 1977 to 1983. The new plan emphasized the power of new genetic technologies to reveal intercrosses and gene flow in the natural history of populations.

The revised policy was intentionally vague and general. It provided the agency officials the flexibility they required to protect species and to prosecute violators who would try to interpret the hybrid policy in their favor. The experience reminded me of the familiar aphorism about the making of sausage and legislation: The process is not always pretty but in some cases the product tastes pretty good.

The political turmoil over the legal ratification of subspecies, inter-crosses, and endangered species represents the tip of the iceberg for a critical but highly contentious conservation issue: When should a subspecies be maintained as pure or when might it be encouraged to intermix with another subspecies? The answer has two tiers. One involves natural gene flow introgression, which we simply observe and document, such as nineteenth-century hybridization between Texas and Florida puma subspecies. The second involves manage-ment intervention to hybridize, as was carried out in the 1996 pan-ther restoration experiment of the same two groups.

Most recognizable flagship species have numerous named sub-species thanks to the eagerness of nineteenth-century mammalogists like Charles Cory and Outram Bangs. In that romantic era, intrepid biological explorers traveled across uncharted terrain and docu-mented specimens by shooting them. Little objective criteria for sub-species distinctions were available, so rather subjective distinctions based on geography and limited samplings were sufficient to define new subspecies and give them names.

Today's taxonomists have better population-based tools and more explicit criteria for recognizing subspecies. And the advent of molec-ular genetics adds a virtually unlimited number of DNA characters to examine. Also, the evolving DNA sequences in species provide an inborn molecular clock that can estimate the time elapsed since two species or subspecies last exchanged genes. Paleontologists tell us it takes one to two million years for one vertebrate species to evolve into another new species. Lions, tigers, leopards, and jaguars are typ-ical noncontroversial species. We can tell them apart easily, their gene differences trace their last common ancestor to about two million years ago, and they do not produce viable hybrid offspring naturally.

By contrast, subspecies designations continually create conflicts in management plans for endangered species. Field ecologists and morphologists who study subspecies note unique characteristics that may be adapted to habitat—the long legs and narrow, flat skull of the Florida panther, for example. These adaptive characteristics are seductive in the taxonomist's yearning to nominate a new taxonomic unit. However, it is difficult to quantify precisely the number and degree of adaptations that render a subspecies truly unique.

Once one satisfies the complexities of naming a subspecies, the tendency is to resist at all costs natural or implemented outcrossing for whatever reason. The ill-fated captive program of the Asiatic lion was a casualty of such "purism." When we discovered that the five founders of the very successful captive Asian lion program included two African lions, the conservation effort was abandoned, much to my chagrin.

A similar conundrum has developed recently with the Amur leopard, a tiny relict subspecies population restricted to the mountains in the Russian Far East near Vladivostok. The subspecies is physically distinctive, and a comprehensive genetic analysis of leopard subspecies by my Russian graduate student Olga Uphyrkina added extensive molecular rigor to the uniqueness of the Amur leopards. A captive Amur leopard population, established in the 1980s, has grown to over two hundred individuals worldwide. However, suspicions that one of four founders of the captive pedigree came from the adjacent North Chinese leopard subspecies now extinct in the wild were recently confirmed by molecular genetic evidence. The wild Amur leopard is reminiscent of Florida panthers with fewer than fifty remaining cats suffering from recent bouts of inbreeding. Will knowledge of the mixed ancestry of the captive pedigree lead to its termination on subspecies "purism" grounds? I am hoping not.

Occasionally I encounter those who defend the opposite extreme. Some biologists and conservationists have concluded that all subspecies are so ephemeral and difficult to discern that they should be ignored completely. Do away with all subspecies and concentrate on species. Facilitate intercrosses between populations, subspecies, and continents to keep genetic diversity maximal.

My own perspective falls in between these diametrically opposed positions. Since subspecies have the potential to evolve into species or at least to acquire habitat-specific adaptations, their protection seems reasonable. Without recognizing these potentials we would not have a reason for protecting the Amur leopard, the Florida panther, the northern spotted owl, or the Asiatic lion. The genetic distinctions demonstrated between subspecies means that two populations have been isolated long enough to change and to adapt. For most subspecies, ones that have been isolated for less than 200,000 years, there is no cogent genetic reason for keeping them separate if they happen by chance to interbreed. However, neither is there any good reason to combine them because mixing would reshuffle the genetic adaptations that have accumulated. The single exception where gene flow should be augmented is illustrated by the Florida panther, where overwhelming genetic, reproductive, medical, and ecological life history data point to imminent extinction. We were right to intervene, but this case was extreme and very much the exception to the rule.

For scientists used to clean solutions, taxonomy is vexing. There is no right way to classify species, subspecies, populations, or other taxonomic hierarchies, only agreed-upon conventions. Scientists, particularly biologists and geneticists, love to challenge established thinking and establish new paradigms. But in taxonomy, this dynamic is largely rhetorical and almost philosophical. Perhaps it is no surprise that Ernst Mayr's most recent books are dedicated to the philosophy of biological thinking.

Conservation biologists use all the scientific tools available to achieve the goals of species and habitat preservation. It seems a shame that less than 1% of all the species that ever lived survive today and that only about 5% of the sum of the world's living species have names. Yet, our preservation efforts must be built on a solid foundation: an ordered taxonomy of living species. So we are forced to do as politicians do—compromise and move forward—often before all the required data are at hand. Every good scientist I know finds such an exercise counterintuitive, difficult, and sometimes impossible. But the really good ones try anyway.

Six

A Whale of a Tale

We coexist on this planet with whales who have a brain not only much
larger, but far more complex and far older than ours. And we don't have a
clue what it is used for.
—ROGER PAYNE

SCOTT BAKER WAS TALL, LIMBER, DEEPLY TANNED, WITH
squared jaw, dark brown hair, and a calm but deliberate drawl cultivated by a childhood in the Deep South. The young man described
unabashedly how he had spent his entire graduate education aboard
schooners traversing the Pacific Ocean from Glacier Bay on the South
Alaskan coast to the Hawaiian archipelago, photographing the black-
and-white tail patterns (he called them "flukes") of humpback whales.
He explained that each whale's fluke is unique, allowing him to recognize individual whales again and again.

I was fascinated as he unraveled the tale of a species driven to the
edge of extinction by commercial whaling, documented in the log-
books of nineteenth-century whalers, of twentieth-century conservation efforts, and of the worldwide demographic computer database
of today's International Whaling Commission. Scott wanted to
explore the nuances of the humpback's population history, and he figured that genetics research might help him. He implored me to equip
him with the genetic artillery that we had adapted to cheetahs, lions,
and pumas.

Humpbacks are glorious creatures, reaching lengths of nineteen meters and weights over forty-eight tons. They inhabit all of the earth's oceans (except the Arctic Ocean) and are well known for their huge pectoral flipper-fins, which when they breach the water look more like the wings of a mythical flying creature. The humpback's affinity for the seacoast made them an easy mark for eighteenth- and nineteenth-century whaling expeditions. Their numbers plummeted from a pre-exploitation high of 125,000 to less than 5,000 before worldwide protection was implemented in 1966. Over the past few decades the population has increased only slightly, so slowly that Scott had concerns that legislative protection might have been too little and too late. He was anxious to investigate the possibility of genetic impoverishment in the hope of helping his beloved whales.

His seemed like a reasonable question and I was intrigued. Yet, I could not help wonder about one small issue: How would one collect DNA specimens from a free-living humpback whale? Scott never flinched.

"Elementary," he explained. "We simply launch a tiny biopsy dart to collect skin specimens suitable for cell culture or for direct extraction of the whale's DNA. I've been doing this for years. In fact, to prepare for the gene hunt, I have already collected samples from over one hundred whales."

He went on to describe a sampling scheme first developed by a wildlife veterinarian named Rick Lambertson. Lambertson had designed a circular dart that fitted on the end of an arrow. Just ahead of the feathers a floatation bobber was also attached. Replacing the arrowhead was a tube-shaped dart with a razor-sharp circular opening at the point. Jutting through the center of the oval was a tiny wire dental barb. The arrow-biopsy dart could be shot at a passing whale using a high-powered crossbow. The dart's circular arrowhead would penetrate the whale's skin and blubber, while the central barb hooked the soft tissue section. The dart tip has a flange collar and is lubricated to fall off the whale when it dives below, releasing the biopsy-laden arrow to float on the ocean surface where it can easily be collected.

Scott was anxious to demonstrate the procedure to me that day. So we went outside to a large field at Fort Detrick, Maryland, where our

National Cancer Institute laboratories had relocated in the early 1980s. The green was adjacent to the army hospital built in the 1950s when Fort Detrick served as a center for offensive biological warfare. We used a cardboard box as a surrogate whale, an eighty-pound crossbow, and five of Lambertson's darts. Four of five attempts netted a cardboard plug; the fifth, shot by me, missed the box!

Scott Baker had trained as a field ecologist at the University of Hawaii, where he studied humpback whale migration in the North Pacific. His research photographs were combined in a vast computer database that matched tail flukes in geographically distant locales, thereby proving the long migrations these animals undertake yearly. Indeed, whales migrate some twelve thousand kilometers from cold summer feeding grounds where they devour huge amounts of planktonic krill, sardine, mackerel, anchovies, and small schooling fishes, to their breeding and birthing grounds in the tropics. During his graduate work, Scott had collected precious tissue biopsies whenever possible. He sampled North Pacific whales feeding off the southeastern Alaska coastland, off central California, on the Baja California coast of Mexico, and from Hawaii.

Scott joined our laboratory in 1989 and over the next few years he became well versed in molecular genetics, DNA fingerprinting, mitochondrial DNA-RFLP, DNA sequencing, and population diversity analysis. Yet, he was perpetually organizing collection trips. The Pacific whale populations had already been extensively sampled, so Scott aimed his crossbow toward the North Atlantic populations.

Tail fluke studies had identified Atlantic humpback whale pods feeding during summer months in the Gulf of Maine, Newfoundland, Greenland, Iceland, and Norway. Every winter the whales from each of these locations would migrate south and coalesce in a treacherous coral reef region called the Silver Bank, just north of the Dominican Republic in the Caribbean Ocean. Scott enlisted Oswaldo Vasquez, leader of the whale researchers at the University of Santo Domingo, to back an expedition for sampling the Atlantic whales when they arrived. He also invited me to come along.

I arrived at Santo Domingo in mid-February. A crowded and stifling four-hour bus ride across the island's steamy rain forest brought me to the tiny village of Samaná on the northern shore. The sea was calm and picturesque. A large and glorious sixty-foot research schooner sent by the Center for Coastal Studies in Provincetown, Massachusetts, had arrived the day before to aid our quest. Scott is not what I would call gregarious, so he did not give me too many details about the adventure to come.

When we boarded the Provincetown ship anchored in the Samaná harbor I presumed we would use it to approach the humpbacks that we could see gallivanting in the bay. I was wrong. Phil Clapham, the Provincetown chieftain, explained that we would navigate a fourteen-foot Zodiac (a military-style rubber pontoon raft), equipped with a twenty-five-horsepower Evinrude outboard motor. The large schooner was to remain anchored comfortably in the harbor.

Scott and I boarded the Zodiac, he manned the crossbow, Phil ran the engine, and Oswaldo was the whale spotter. As Baker strapped himself to the boat with a hemp tether, I asked if we should use life vests. He replied dryly, "No, that would only prolong the agony."

We motored into Samaná Bay and searched for the humpback's signature, a respiratory explosion that invariably inspires someone to cry, "Thar she blows!" We headed our craft toward the first sighting, slowly approaching the blow spot long since vacated by a diving humpback whale. Shortly, a whale that seemed the size of a Boeing 727 broke the surface and gave us a tail wave. When he surfaced the next time, not quite so dramatically, Baker delivered the arrow to the surfacing hump for a direct hit. The whale reacted like he had been stung by a bee and dove rapidly below. We recovered the dart, moved away and started searching for our next blow.

A female being pursued by a few suitors provided our second opportunity. Our vessel bobbed across six-foot whitecaps as we jock-eyed to avoid being upended between a pair of fifty-ton humpbacks hell-bent on copulation. The earnest males seemed annoyed to be competing with a noisy motorboat for their lover's attention and showed their irritation by swimming rather close. When one actually lifted the Zodiac above the waves playfully, I began to wonder if this civil servant might be better off on terra firma or at least on the mother ship docked in Samaná Bay. Unfazed by the turmoil, Scott popped both males and collected the biopsies.

By the end of the collections, my terror began to subside. The humpback whales impressed me as being very gentle, albeit remarkably large, particularly when viewed from a tiny craft at eye level. The darts did not seem to faze them much and the sight of six or eight humpbacks gyrating playfully in a bay not a half-mile offshore was unforgettable. Their tragic history and thorny genetic questions would later come into sharper focus, but by the end of our visit to the Silver Bank, I truly understood what drove Scott's burning passion for preserving this majestic evolutionary creation.

Vicariously through Scott Baker I would learn a great deal about whales and their conservation status. Some fourteen families and eighty-three species of the mammal order Cetacea (whales, dolphins, and porpoises) roam the world's oceans. Cetaceans are generally split into two major categories or suborders. The first, baleen whales (Mysticeti), are named for their baleen whalebone plates, filter feeders that sieve out small planktonic prey. The second group comprises the toothed whales (Odonticeti), which includes sperm whales, beaked whales, dolphins, and porpoises. Several species, notably blue whales, fin whales, right whales, gray whales, and humpbacks, have been heavily exploited by the whaling industry and have only recently been afforded international protection.

Over the years Scott had amassed biopsies from four North Atlantic humpback populations (Maine, Newfoundland, Iceland, and the Silver Bank), four North Pacific populations (Alaska, California, Mexico, and Hawaii) and five populations from the southern oceans (Antarctica, West Australia, East Australia, New Zealand, and Tonga). Using minisatellite DNA fingerprints, mitochondrial RFLPs, and control-region DNA sequences, he examined the molecular genetic diversity in each pod. Remarkably, all the populations still displayed appreciable genetic diversity. In spite of extreme exploitation, the humpbacks had not yet suffered prolonged or repeated episodes of inbreeding such as had afflicted the Gir lions or the Florida panthers.

That was indeed a fortunate break for the whales. They had survived the heyday of the whaling assault without notably compromising the humpback species genetic potential. So far, so good.

Scott and I did notice some intriguing patterns in the genetic variation that gave us insight into several mysteries surrounding the whale's migratory habits. Humpback watchers knew that whales from several geographically distant feeding grounds would come together at shared winter breeding regions (Hawaii in the Pacific and the Caribbean Silver Bank in the Atlantic), but then what happened? Was there a homing strategy or did they mix up their destinations on return trips?

The genetic data had the answer. Whales from each feeding ground had a characteristic group of unique genetic types, a population "signature" pattern that was recognizable and unique to the locale. For example, the mitochondrial genetic types (genotypes) from each feeding ground were particularly distinctive. Remembering that mitochondrial DNA is inherited from mothers only, we were able to conclude that the migration back to summer feeding grounds must be determined by female leadership. If whales were not returning faithfully to their mother's original location, each feeding ground would have contained a mixture of genotypes from several other feeding locales.

Humpbacks from Alaska, California, and Mexico migrate to Hawaii annually to mate and give birth. Mixtures of all their genotypes turn up in Hawaii every year. Newborns then follow mom back to her original feeding site. The genotype distribution also indicates that males from each locale preferentially mate with females from their own region, even though other opportunities are available. No one is certain how this occurs, but assortive mating, the choosing of mating partners from their natal region over mates from different geographical locales, is clearly happening in these whale populations.

The humpback's molecular genotypes were also informative when the populations in different oceans were compared. We analyzed individual whales' mitochondria and microsatellite DNA genotypes using a computer routine that builds an evolution-based phylogeny— a branching treelike diagram that connects each whale to others based on the similarity of their genes. For example, a phylogenetic tree of primate species would connect a human and chimpanzee genotype as close relatives, then connect that pair to orangutan's (a more divergent great ape), then that group to the gibbon's and siamang's—the lesser apes—then all the apes to more divergent Old World monkeys like baboons or green monkeys. Building such phylogenetic trees with genetic and DNA sequence data from individuals, populations, or species provides a powerful tool we use to interpret historic separations, migrations, connections, and evolutionary hierarchies among different groups.

When Scott analyzed whales from all over the world he found three major phylogenetic clusters, or oceanic clades (a clade is a group of phylogeny-based similar genotypes). Not surprisingly, the whales in the southern oceans were more closely related to each other than to whales found in the North Pacific or the North Atlantic. This is what we would expect if gene flow or migration between these regions had been absent or highly restricted. Actually, nested within the three oceanic clades of humpbacks were four exceptional "miniclades." These were small groups of three to five closely related genotypes found in one ocean that as a group closely resemble the major clade in another (e.g., a small group in the North Atlantic that more closely resembles the southern clade). This pattern is best explained by an ancient very rare migratory event between the oceans. That migration established a new lineage that in the present day appears to be in the "wrong" ocean.

The quantity of genome diversity in all humpbacks also allowed us, via the molecular clock, to estimate the time elapsed since the species last passed through a population bottleneck. Remember we used this approach to date bottlenecks in African cheetahs and North American pumas to 12,000 years ago. The humpbacks' genetic variation reached back to some time between three and five million years ago, so the whales have been free of significant genetic homogenization or population bottlenecks for a very long time. For comparison, similar genetic estimates date modern Caucasian or Asian human populations' descent from a migration "out of Africa" between 150,000 and 200,000 years ago.

Although the three-to-five-million-year interval that built the humpback's genetic variation would bode well for the species, the scenario also has a dark side. When an outbred healthy species persists for a very long time, a lot of potentially damaging mutational gene variants accumulate in the population. The mutational accumulation at many different genes over long periods of time, called the "genetic load," is protected from natural selection by the diploid state. That is, as for humans, whales, and all vertebrate species, each individual has two copies of each gene, one from each parent. In a healthy, outbred population the normal gene mixing of reproduction makes the pairing of two damaged genes, the resulting expression of the mutation, very unlikely. The genetic load shows its face when populations drop to numbers so low as to promote inbreeding. We have to believe that four million years of bottleneck-free subsistence has allowed an enormous genetic load, a genomic vulnerability, to accumulate in humpback whales. Fortunately, the hidden mutations have not yet been uncorked. It appears that the rapacious slaughter of this species was suspended none too soon.

Whaling as a commercial endeavor dates back to twelfth-century Basque sailors in the Bay of Biscay off the northern Spanish coast. Harvesting whales for food, blubber, and oil demanded considerable skill, nautical organization, and dexterity with handheld harpoon equipment. Modern whaling is said to have begun in 1864 when Svend Foyn, a Norwegian mariner, introduced a harpoon gun that fired from the bow of a small steam catcher. By the 1930s some 30,000 great whales were being harvested annually, with commercial whaling dominated by Norwegian, British, and U.S. whaling companies. As early as 1911, conservationists at the British Museum of Natural History began calling for scientific monitoring of the intense and unrestricted slaughter of the great whales, particularly the humpbacks.

The International Whaling Commission (IWC) was established in 1946 by forty whaling nations ostensibly to achieve maximum sustainable utilization of whale populations and by default to ensure the future of whale stocks as a harvestable resource. That body remains today as the primary watchdog for whale conservation and the whale industry. By 1962, humpback whale numbers, once as high as 125,000 worldwide, were so reduced (5,000–10,000) that the industry collapsed. The next year, whalers worldwide agreed to stop hunting humpbacks in the hopes they would recover. By 1982, the IWC voted for an indefinite global moratorium on all commercial whaling; it became effective in January 1986. In addition, the Convention on International Trade in Endangered Species (CITES), an international conservation treaty, listed all the great whale species regulated by the IWC in Appendix I, the most endangered category. This means that all CITES signatory nations agreed to ban international trade on the species and their products.

Despite strong international criticism, four countries, Japan, Norway, Iceland, and South Korea, took exception to the IWC and its whaling moratorium, citing the 1946 International Whaling Convention, which established the IWC and explicitly permitted the taking of whales for scientific purposes. Japan also filed objections to the CITES listing for six whale species (fin, sei, Bryde's, minke, sperm, and Baird's beaked whale), claiming they were not really endangered or threatened.

The Japanese whaling industry established the Institute for Cetacean Research (ICR) in Tokyo as the scientific arm of Japan's whaling program. ICR has been taking 400 minke whales per year from Antarctica since 1987 and about 100 minke whales from the North Pacific since 1994. By 2000, they added 50 Bryde's and 10 sperm whales to the list.

The ICR's so-called scientific whaling is technically legal since any nation's compliance to IWC and CITES is voluntary. However, the bulk of the take is destined for retailing as *kujira,* generic for whale meat, in Japanese fish markets. In fact, half of the ICR's seventy-three-million-dollar annual budget is recouped from sales of whale products. Japanese ICR scientists argue that the research is necessary to gather important data on stock management, and they complain that their critics ignore the fact that minke whales are rather plentiful and not particularly endangered. Their latest ploy, presented at the

July 2001 meeting of the International Whaling Commission, was to suggest that whales eat too much and should be blamed for decreasing ocean fish stocks.

Most IWC member nations have not been impressed with either the quality of the new scientific data or the arguments marshaled by the Japanese ICR to defend their take. The U.S. delegation called for international sanctions (unsuccessfully), and in one of his last directives in 2000, President Bill Clinton banned Japanese whalers from U.S. waters to register symbolically an American objection to Japan's refusal to honor the IWC ban.

As this debate heated up, Scott Baker nursed another concern. He had always wondered why humpbacks and other threatened species stocks were so slow to recover. After all, hunting of humpbacks had been suspended for decades. His genetic data showed surviving humpbacks had considerable residual genomic diversity. So what was the problem? Why hadn't they recovered their numbers as had other protected species like alligators, northern elephant seals, or American eagles?

Scott suspected that illegal hunting, perhaps under the guise of legitimate or accepted legal hunting, was responsible. In the early 1990s, Scott moved to the University of Hawaii to work with evolutionary geneticist Stephen Palumbi, and the two of them conjured up a scheme to get an answer to this question.

Scott had collected tissue specimens from hundreds of whales, including all the great whale species and other smaller cetaceans. He extracted DNA sequences from mitochondrial and nuclear genes while searching for the evolutionary hierarchy or phylogeny of the whales' natural history, and in the process he was able to derive signature DNA sequences that identified different whale species unambiguously. Empowered by this evolutionary database, Scott and Palumbi set out on a covert but noble mission, one that would change forever the international monitoring of whale harvests.

Rumors of endangered whale species being harvested under the Japanese "scientific whaling" loophole were rampant in conservation circles. Nonprofits like Greenpeace, Earthtrust, and the International Fund for Animal Welfare were vocal in their suspicion that all Japanese sushi and kujira were not what they seemed. Scott and Palumbi had the scientific tools to find out. They ran some pilot experiments to see whether they could detect whale DNA from samples of raw fish meat, or sushi. They could, so in 1993, Scott headed incognito for Tokyo, where for the first (of many) times he visited the city's bustling fish markets.

A six-foot-three Caucasian American with an Alabama accent snooping through a Japanese fish market might have attracted attention, so Scott asked a native Japanese conservationist, Naoko Funahashi, to go to different markets to purchase fresh whale meat, anything labeled "kujira." Funahashi traveled first to the Tsukiji fish market in Tokyo, the largest in the world, a fifty-six-acre jumble of stalls that retails five million pounds of seafood every day. There were over four hundred marine species offered. She stuffed whale meat chunks in small glass vials, labeled with the advertised name and the market location. Then she slipped to smaller out-of-the-way markets to add to the undercover tissue samples.

Scott improvised a makeshift genetics laboratory in his Tokyo hotel room. He needed to process the whale tissues in Japan because legally he could not take them out of the country. His goal was to prove that at least some of the whale meat had come from CITES Appendix I endangered species, and to move such tissue specimens or even DNA across an international border would be illegal unless they were accompanied by a Japan-endorsed CITES export permit. His chance of getting one was flat zero, given the motive for his research.

Instead Scott used a new technology called "polymerase chain reaction" (PCR), an enzymatic DNA photocopy reaction, to make synthetic copies of genes extracted from the kujira samples. After the PCR-DNA photocopying, Scott separated the copied DNA stretches from the original whale DNA template using an electrophoresis gel rig. The synthetic DNA was an exact likeness of the whale meat samples, but, like a photograph of an endangered species, it would not be regulated by international legal prohibition of transport between countries. The original whale DNA was left behind, so no laws were broken.

Scott performed the DNA amplification in a packed hotel room, atop a PCR minibar, surreptitiously and deliberately. Once he had gathered several dozen whale meat specimens, he strolled nonchalantly through customs, with his mobile lab and DNA copies secure in his luggage. He boarded the plane home anxious to sequence the DNA photocopies, which would reveal the Japanese whale meat's true identity.

The results of his sequence analysis were definitive and damning. In sixteen initial specimens purchased, he identified four endangered fin whales, one humpback, and several minkes, dolphins, and porpoises. Clearly, protected species like fin and humpback whales were making it into Japanese fish markets.

Over the next several years, Scott visited Japan several times and then he expanded his surveillance to the markets of South Korea. He continued his work with the assistance of several Ph.D. students through the 1990s from his new position as senior lecturer at the University of Auckland in New Zealand. His team collected over seven hundred kujira specimens from markets throughout Japan and three hundred from South Korea. DNA from each was amplified by PCR in locked hotel rooms, the synthetic DNA products hand-carried overseas, and the species identified by DNA sequence technology back home.

Between 1993 and 2000, fully 10% of the retail whale meat samples typed out as prohibited whale species. Japanese officials had insisted that these species were not being taken by their scientific whaling program. In all, Scott identified six species of baleen whales: 24 fin whales, 5 sei whales, 2 humpbacks, 4 Bryde's whales, 2 blue whales, 1 Asian gray whale, and 1 blue/fin hybrid. While most of the take were in fact minke whales, which were not particularly endangered, even these held a surprise. Nearly one-third of the minke whale samples were shown by phylogeny analysis to originate from the Sea of Japan, where a small endangered population of minke whales is supposed to be legally protected from whale harvest altogether. Minkes from the more plentiful Antarctic and North Pacific population represented only two-thirds of the specimens.

Scott's screen of South Korean fish markets did not fare much better. Bryde's whale, beaked whale, humpback whale, killer whale, and dolphins were discovered masquerading as legal whale meat. Over several years, Scott and Palumbi published a stream of scientific reports detailing their forensic surveillance. They pleaded with the IWC, and with the Japanese and South Korean governments, to adopt their molecular genetic methods to screen their catches and to enforce their agreed-upon protection of the endangered whales. In 1995, the IWC adopted a resolution to implement a program for spot testing of whale meat using DNA identification. Japanese wildlife officials have also agreed to monitor their catch using Scott's methods.

The macabre moral of their genetic detective work was that Japan and South Korea's "legal" exception to the IWC whaling moratorium had for decades provided an effective cover for the illegal harvesting of protected whale species. The refusal of Japan, South Korea, Iceland, and Norway to comply with the 1986 IWC whale-hunting moratorium has made these nations pariahs in the international marine mammal conservation community, but it turns out they are not alone.

In 1994, the conservation community was rocked to learn from a group of Russian scientists that Soviet factory ships harvested as many as 48,000 humpbacks in the southern oceans between the 1940s and the late 1970s. Their official government reported catch during that period was 2,710. The latest IWC records suggest that 80–95% of the humpback's original Southern Hemisphere population was hunted illegally during those years.

It now seems certain that illegal catches heading for Asian fish markets have significantly limited the humpback whale's recovery from their historic slaughter. Finally recommendations about whale meat monitors are being taken seriously and further international cooperation may plug a loophole that exploiters of wildlife have enjoyed for decades. Genetic technology made a difference in this case, and may do even more in the future.

The insight that Scott Baker and his collaborators have gained is truly remarkable. These pioneers worked out a way to study the migration behavior of a species that shows itself only briefly when it crests out of the water. Today the accumulated knowledge of the humpback's migratory patterns, maternal dominance, natal range fidelity, and navigational skills is considerable. In many ways we understand more about the humpback whale's modus operandi than we do about more familiar species of land animals. We should hope one day to understand as much about aardvarks, skunks, bush pigs, or giraffes.

The integration of available biotechnology for monitoring whales, for tracking their genes, for identifying individuals, and for assessing their history has made these discoveries possible and has changed forever the "scientific whaling" debate. The knowledge derived is not just an academic exercise; rather, it provides the managers and conservationists of whales with crucial information. Nothing can topple a clever, well-articulated, firmly rooted opinion like cool, hard new scientific data that contradict it. The collecting of forensic results took years of patience, determination, and visionary acumen. When those qualities are married with science research and policy, they really can change the world for the better, as Scott Baker has done for his beloved humpbacks.

Seven

The Lion Plague

THEY AROSE AT FOUR A.M. IN EAGER ANTICIPATION OF THE high point of their African safari, a hot-air-balloon ride across the vast Serengeti topped off with a champagne breakfast. Connie and Harold Chandler eagerly donned their Banana Republic vestments, then hurried to the balloon launch pad, a short Land Rover ride from their luxurious Seronera tourist lodge. It was January 1994, the perfect season to view the expansive vista of East Africa and its marvelous wildlife. The evening before their tour drivers had regaled them with apocryphal yarns of obliging Robert Redford, Meryl Streep, and their crew during the filming of *Out of Africa*. The Chandlers gathered snacks, sunscreen, and their camcorder onto the balloon deck in anticipation of a splendid Serengeti adventure.

As the sun crept up over the sweeping Serengeti Plain, the balloon lifted above the endless herds of wildebeest, zebra, and giraffe who had grazed the region to a brown stubble. Those fortunate enough to have visited an East African game park would spend countless hours in animated conversation reliving their euphoria. Most would also agree that even the best photographs are woefully inadequate in expressing the grandeur. You just have to experience it to understand the extraordinary enthusiasm the savannah inspires.

The Chandlers were flying actually and spiritually when Connie spotted a trio of young male lions strolling below. The balloon descended and Harold started the camcorder rolling. Then it happened.

The male in the rear began to twitch his whiskers. Soon his connniptions became noticeable enough to spook his brothers, who took a few swats at the trembling adolescent and ran off, probably to avoid the descending balloon. The shivers progressed to tremors and then to severe muscle spasms. Then the unfortunate beast lurched into a grand mal neurological seizure, extending his front legs forcefully, flailing him into the air and crashing to the ground. For thirty agonizing minutes, the animal thrashed about as if possessed by demons. Finally as the violent seizures and gyrations gradually lost intensity, the miserable creature breathed his final torturous breath before the horrified ballooners.

No one felt much like drinking champagne.

The shaken tourists had witnessed something that few people, even trained veterinary clinicians, had ever seen: the acute neurological collapse of a large predator. But what killed the lion? Was it a poison? Was it a nasty germ, or was it a hereditary ailment?

Greg Russell, the visibly upset balloon pilot, went looking for the one person in Tanzania who might be able to diagnose the malady, the park's new wildlife veterinarian and my old friend, Melody Roelke.

Not long after the incident, I received a cable from Roelke that immediately caught my attention. Her note was short, cryptic, and troubling.

"Steve, please call me. We have a situation! The lions are dying of a mysterious disease. I am worried that a fatal version of FIV is running through the prides. Help. Hurry. Luv, Mel."

Melody then sent me a copy of the Chandlers' video. There was no time to waste. I briefed Janice Martenson, my scientific first lieutenant. She phoned Craig Packer, the ecologist who was leading a twenty-year project on the Serengeti lions, and Linda Munsen, a savvy medical pathologist from the University of Tennessee with extensive wildlife experience in emerging feline infections. Then we contacted Max Appel at Cornell and Hans Lutz in Zurich, crack virologists who would recognize old and new cat viruses and who knew the frightening FIV viruses very well. These represented the best minds in the business to uncover the source of the lion's plague. All were anxious to help.

Several years before the balloon ride, the cat family Felidae had injected itself abruptly into mainstream AIDS research when Marlo Brown, a private domestic cat breeder in Petaluma, California, concluded that one of her pet cats was dying of AIDS. It certainly had symptoms like a human AIDS patient: severe weight loss, respiratory infections, skin lesions, and numerous bacterial infections. Even though these symptoms were hallmarks of immune suppression, the idea that the cat had AIDS seemed ludicrous.

AIDS had first appeared in people in the early 1980s in a clustering of patients from homosexual communities in Los Angeles and New York City with a rare cancer, Kaposi's sarcoma, and pneumonia. It was later determined that AIDS patients suffer a gradual loss of a certain type of lymphocyte, the CD4-bearing T-lymphocyte, an important player in immune defense against viral diseases. AIDS is caused by a new human virus, termed "human immunodeficiency virus," or HIV, which infects and destroys CD4-bearing T-cells. HIV appears to have first entered humans early in the twentieth century, originating from an African primate species infected with simian immunodeficiency virus (SIV), HIV's closest genetic relative.

When Marlo Brown's sick cat was presented to Dr. Niels Pedersen, a seasoned animal virologist from the University of California at Davis, HIV and SIV were the only viruses of this type known. Pedersen was curious about the possibility that cats might have their own AIDS virus. He knew that cats harbor many other viruses, such as the feline leukemia virus and the feline infectious peritonitis virus, which had afflicted the cheetahs. So he tried a standard cell culture virus-isolation technique from blood samples of Brown's pet cat. Sure enough, he was able to isolate a feline version of the AIDS virus, which he named "feline immunodeficiency virus," or FIV.

FIV has a genome similar in sequence, in gene content, and in gene arrangement to that of HIV. FIV and HIV are lentiviruses, a subfamily of the retroviruses, agents that, like the Lake Casitas mouse virus, can cause leukemia or sarcoma in mice, cats, monkeys, and chickens. Like HIV, FIV has a small viral RNA genome of about nine thousand nucleotide letters, an average size for retroviruses. HIV and FIV both contain the four major genes common to retroviruses: *pol,* which encodes the reverse transcriptase enzymes that convert RNA genome to a DNA copy suitable for slipping into a cell's chromosome; *gag,* which specifies viral core proteins to protect and shield the sensitive RNA genome; *env,* the outer envelop proteins of the virus that bind to cell surface receptors to trigger virus injection into a cell; and *LTR* regulatory on/off gene rheostat sequences at both ends of the virus genome.

Once Pedersen published his discovery of FIV, virologists began to screen for FIV across the veterinary research community. Varying by region, 1–10% of domestic cats sampled throughout the world were infected with FIV. FIV was shown to cause a gradual depletion of CD4 T-lymphocytes, the same cells destroyed by HIV in AIDS patients. FIV-infected cats displayed many AIDS-like symptoms, including an early flulike syndrome, uncommon cancers, lymphomas, neurological tumors, upper respiratory disease, and bacterial infections. The cat AIDS epidemic and its causal virus were carbon copies of the human AIDS epidemic.

The isolation of FIV stimulated an intense research effort focused on FIV in cats. A few years after Pedersen's announcement, I attended a Washington, D.C. conference on newly discovered viruses in humans and animals. Bob Olmsted, a young postdoctoral fellow at NIH, summarized the latest estimates of FIV prevalence in the world's house cats and presented a litany of the varied diseases that FIV-mediated immune collapse would induce. The upshot was that FIV posed an insidious scourge to cats, just as HIV did to humankind.

At the coffee break, Bob and I wondered aloud: What if FIV gets into other species of cats? Or had it already moved to wildcats as HIV had jumped from monkeys to humans? There are thirty-seven species of cats in the Felidae family, and of these, all but domestic cats are considered threatened or endangered. Would FIV finish off some or all of them?

What about the species with a history of genetic compromise, left over from population bottlenecks and close inbreeding? Domestic cats are rather outbred, but cheetahs, Asiatic lions, and Florida panthers all carry a genetic hangover from narrow escapes from extinction. I explained that these species had lost considerable genetic diversity during recent population crashes. Their genetic monotony would render their immune system a vulnerable target for new viruses.

Bob agreed to help us screen my collection of frozen wildcat serum specimens for antibodies to FIV. Fifteen years of adventures in wildlife genetics and reproduction had produced a treasure trove of tissue samples collected from thirty cat species. A serum sample will retain antibodies to any microbe that had infected the cat, and we had thousands of samples stored in my freezers that had never been examined for FIV.

Bob would use a technique called western blot electrophoresis to screen the serum. The procedure exposes disrupted proteins from purified domestic cat FIV, which Niels Pedersen had sent us, to serum from our wildcats. If the big cat was infected with FIV or a relative of FIV, antibodies in the cat serum would bind to FIV proteins. The complex of antibody bound to the FIV core and envelope proteins could then be separated and easily visualized on a western blot electrophoretic gel.

We were taken aback by what we found. Virtually every cat species we examined had some individuals with FIV antibodies. Lions, tigers, cheetahs, snow leopards, ocelots, pumas, all of them! If we had more than twenty serum samples from a species, then at least a few animals tested positive for FIV antibodies. Detection of antibodies meant the cats had been exposed to and were probably still infected with a virus identical to or closely related to FIV. In all, nearly 3,000 cat sera were screened. Of 434 pumas collected from throughout their natural range in the Americas, 97 (22%) had FIV antibodies. The Florida panthers had a 24% infection rate. African lions were even worse. Over 500 of 700 tested African lions were positive for FIV antibodies. The legendary Serengeti lions had an incidence of 70%. Serengeti lions older than three years showed a prevalence of 100%. Ocelots in South America tested positive 12% of the time. Our list went on and on. I was terrified that all these cats were on the verge of immune collapse, caused by a deadly AIDS virus.

I sent out an alert to field biologists and zoo lion curators. FIV screenings were repeated in zoos, veterinary schools, and wildlife reserves across the world. Zoo veterinarians were poised to spot the wasting disease, bacterial infections, lymphomas, and neurological destruction described in human and domestic cat victims. The community looked, but prayed that they would not see the realization of our fears.

It took years to assuage my angst. Slowly the negative data trickled in. The Serengeti lions had no apparent immune suppression, no illness, no mortality, and no telltale symptoms. Years of field observations proved that lions and pumas were living to ripe old ages in spite of their infection with a virus that confers a death sentence on house cats. Even FIV-infected cheetahs and Florida panthers, as genetically distressed as they were, were not dying of immune system collapse.

Wildlife veterinarians searched and examined hundreds of FIV-infected animals in zoos, but they could not come up with a disease to connect with FIV infection. By the mid-1990s, I was ready to conclude, albeit cautiously, that although FIV would effectively kill domestic cats, it was not doing the same in free-ranging species. The wildcats seemed to be susceptible to infection, but somehow immune to the fatal disease.

So how did they avoid AIDS? What do these big cats have that their domesticated cousins and humans do not?

By then, my students and I were hooked on the puzzle and determined to solve it. We began by examining the patterns of genetic variation among FIV in the big cats. We hoped reconstructing the natural history of the virus would provide some clues. Eric Brown was an easy sell for the project. A determined, impassioned, and clever graduate student, Eric took the FIV mystery to the next level by tracking the evolutionary patterns of the virus's genome sequence from lions collected across Africa.

Eric discovered that each infected lion was producing a swarm of FIV genetic variants, even when he sampled the most slowly evolving (or mutating) FIV gene, *pol,* FIV replicates in a number of discrete tissues in an infected lion, producing several hundred million new viral particles daily. A high mutation rate contributes one or two new nucleotide changes in each new particle, so when Eric determined genome sequences from a dozen virus particles in a single lion, each one differed in several nucleotide letters from the others, though they were still similar enough to reflect their descent from the virus that first infected the lion.

In effect, the FIV infection produced a rich diversity of virus flavors. The FIV was flooding the infected animal with millions of distinctive immunological challenges—a recipe for overload of host defenses. To dispatch the infection the host's immune system must recognize and destroy hundreds of millions of new mutational variants each day. This is precisely the enormous challenge HIV presents to AIDS patients and the researchers trying to find treatments. It is no wonder that HIV ultimately overwhelms human immune defenses. FIV was using the same strategy in infected lions. So why didn't these lions die?

Each individual in the Serengeti lion study group had its own slightly different genetic swarm of FIV genomes, indicative of the dynamic change within and specific to each lion. However, when Eric compared FIV sequences of lions from several different populations across Africa he discovered three rather divergent FIV groups—A, B, and C strains. On average, 23% of the *pol* gene nucleotide letters compared among FIV A, B, or C strains were different. FIV *pol* sequences in a single lion differed by only 1–2%, and among lions infected with the same strain, the greatest difference was 5%. The large difference between strains probably means that the virus had evolved in three geographically isolated lion populations, or perhaps even in a different cat species like a leopard or a caracal. Recently, the three FIV strains somehow got mixed together in the Serengeti lion population.

We also discovered that some lions were infected with more than one strain of FIV. Multi-strain infections would give the virus ample opportunity for genetic recombination. Simply put, the different virus strains can swap genes within a lion to create a stronger, hotter strain. This established a serious potential for changes in virulence. Still, as hard as we looked, the lions in Africa were not showing any signs of FIV-induced disease.

The diversity patterns in lions should be viewed in the context of the FIV situation in other felid species. Each wildcat species has its own strain of FIV, only distantly related to FIV isolated from other cat species. So the viruses from domestic cats each had as their closest relatives FIV from other domestic cats. FIV from lions were all closer to other lion viruses than to FIV isolated from pumas or leopards. The absence of any genetic mixing of viral strains among different cat species indicates that once FIV gets into one species it stays with that species. So lion FIV adapts, evolves, and transmits among lions, but rarely, if ever, would jump to another species. This may not have always been the case, but it is true now.

The next key to the puzzle came from a study of pumas, undertaken by a bright and determined New Zealand–born postdoc named Margaret Carpenter. Margaret would examine our expansive stash of serum samples from thirty populations of American pumas. That collection, a bonus from the puma gathering expeditions of Melanie Culver, Warren Johnson, and Melody Roelke, totaled over four hundred individuals and extended from northern British Columbia, south through the Rocky Mountains, Central America, Brazil, and Argentina, to the southern Chilean tip of Tierra del Fuego. Margaret's Pan-American puma FIV sequences showed even more sequence divergence than Eric saw among African lions. She found fifteen distinctive puma FIV strains, each of which was comparable or of greater divergence than the three lion strains. The amount of puma-FIV diversity was enormous, greater than in any other feline species, greater than HIV variation in humans and SIV variation in monkeys, and greater even than the difference between SIV and HIV genomic sequences.

Our obsession with quantifying the amount of FIV genome variation from different cat species was not just an intellectual dalliance. Geneticists know that mutational variation accumulates in a time-dependent manner. So the quantity of FIV variation can tell us how long a virus has been in an individual, in a population, or in a species. Domestic cat FIVs sampled across the globe showed about an 8% difference between any two FIV isolates we compared. African lion FIVs have an average genetic difference of 16% and pumas 18%. These values mean that the puma and lion viruses are much older than the less diverse FIVs of the house cat. These quantitative estimates were just what we needed to piece together a plausible history for FIV in cats. But first I need to explain the typical fate of a species that becomes infected with a murderous infectious agent.

Put simply, when a fatal virus enters a new species, either the epidemic drives the species to extinction or the two learn to live together. Plenty of species have gone extinct in the history of mammals and a great many were victims of a deadly viral disease.

So how do the survivors avoid this fate? Two sorts of events could lead to species survival. The first would involve genetic changes in the virus that temper its virulence. Such a virus becomes attenuated or ameliorated with respect to disease induction and the species lives on.

The second survival path depends on the natural genetic diversity in the afflicted population. Some lucky few individuals in an outbred population are able to employ natural genetic variants in immune defenses to resist the new virus. With time the sensitive individuals perish, while the resistant ones survive, passing their genetic resistance to their offspring. After a few generations, the virus may continue to spread to new individuals, but it kills few, if any, of the new "naturally selected" resistant generation.

Both of these scenarios would lead to a standoff between virus and host. The virus would persist, but not cause anywhere near the devastation its forebears wrought when first infecting the species. We call the change in the virus, in the host, or in both that leads to the standoff an adaptive episode. When scientists encounter a population after an adaptive episode, they witness a delicate balance between a resilient host species and an ostensibly innocuous virus, bearing peaceful witness to a past genomic battle.

Now consider the cats. FIVs in pumas and lions have enormous intrinsic genome diversity, indicating that FIV has been in the wild-cats for a very long time. The viruses in each species are genetically specific for that species, indicating that they got in and stayed put. Most, if not all, wildcats have passed through and survived a historic adaptive episode leading to today's delicate standoff.

By contrast, the domestic cat's FIV has much less variation than that of the wildcats' FIV, and house cats do get AIDS. Domestic cats are simply the most recent unfortunate species to acquire FIV. Their adaptive episode is still in an early stage and will someday lead to selective survival of genetically resistant individuals or to weaker attenuated FIV strains. (Lacking a vaccine or cure, HIV/AIDS would surely follow the same course in human populations. But that is a discussion for later in the book.)

A tantalizing corollary to the adaptive episode hypothesis might help explain the persistence of benign FIV infections in lions and pumas. Retroviruses like FIV do two things regularly. First, when they infect a tissue they trigger a phenomenon called "viral interference." The infection of a cell by one virus will block a secondary infection by a related virus. The Lake Casitas mouse, chronicled in Chapter 1, revealed a good example of viral interference. The mouse acquired a viral *env* gene that produced a protein that sat upon the cell's receptor portal, much as a virus would, thereby protecting the mice from infection by the actual, deadly virus.

Second, FIV infections stimulate antibodies and a more sophisticated T-cell-mediated immunity to dispatch the retrovirus. Could it be that today's FIV-infected Serengeti lions are enjoying a natural FIV vaccine? Could the benign FIV immunize the lion and exploit viral interference to protect lions from a virulent hot strain of FIV that could develop any day?

Unfortunately, we are not able to estimate the precise age of the feline and primate versions of the AIDS virus because viruses do not leave fossils that we can use to calibrate a virus molecular clock with actual years. The viruses could be millions of years old, but a better guess is 10,000 to 100,000 years. We believe they are on the order of 10,000 or fewer years because lentiviruses never appear as endogenous viral sequences in their hosts' chromosomes. Older retroviruses invariably leave their endogenous footprints in the genomes of species they afflict.

Lentiviruses have been isolated so far from horses, goats, sheep, cattle, cats, monkeys, and humans. Genome sequence analysis of all these viruses together suggest that FIV is a bit older than HIV or SIV. It turns out that FIV viruses from felids and SIV from primates are more closely related to each other than either is to the cattle, sheep, or goat lentiviruses. When we considered the accumulating pieces of this puzzle, it became possible to imagine a hypothetical but intriguing history for the origin of these viruses.

Thousands of years ago, around the Mediterranean Basin, a cat attacked its prey, a large bovid ancestor infected with bovine immunodeficiency virus (BIV). The bloody exchange allowed a mutational variant of the bovine virus to infect the large cat and begin a new epidemic in the cats. The virus spread throughout the species, becoming FIV, and the new FIV jumped occasionally to other cat species. The new virus probably killed thousands, even millions, of cats. It may have eliminated the notorious saber-toothed tigers of North and South America that went extinct so abruptly in evolutionary terms around twelve thousand years ago.

With time, adaptive episodes among the infected cat species led to a balanced truce, a commensalism. After that, but still many thousands of years ago, an FIV-infected cat attacked a small African monkey and transmitted an FIV variant that successfully infected the monkey. SIV developed and spread to Africa through the original and then several other monkey species, one of which eventually passed a variant on to humans.

The evidence for the tale is circumstantial, yet the story makes sense when we consider viral diversity, viral evolutionary relationships, disease incidence, and other epidemiological observations. All told, the cats seemed safe from annihilation by FIV, or so we thought until an American tourist couple named Connie and Harold Chandler took a fateful balloon ride in a faraway wildlife reserve in East Africa.

Melody Roelke was worried and stressed. She was finding more and more listless, emaciated lions wandering about in a season of rich prey availability. Neurological symptoms and a wasting syndrome were both warning signs of an immune deficiency. And the video from the balloon ride proved the lions were under attack from a deadly agent. Had a virulent variant of FIV abruptly arisen in the lions? A week after Roelke's cable arrived, Reuters broke the story on its wire: the famous Serengeti lions, home of the Lion King, were dying of AIDS.

But that news report turned out to be erroneous. It wasn't AIDS at all.

Roelke and her Tanzanian veterinary team set into motion the wildlife equivalent of an international public health alert. Dawn to dusk her battered Land Rover went from lion pride to lion pride. The team observed, anesthetized, listened to heartbeats, drew blood, and collected tissue specimens from dying lions. Tissue samples from over a hundred lions were shipped to NIH, to Zurich, to Cornell, and to the University of Tennessee with the hope of narrowing the cause of the plague. Lion demographer Craig Packer produced the horrific estimation that within eight months over a thousand lions, one-third of the Serengeti lion population, had perished.

Several dying young lions tested free of FIV, excluding that as the culprit. The agent did turn out to be a virus, but one we never expected. Autopsy lesions in brain sections had an eerie semblance to crystal deposits seen in dogs infected with canine distemper virus (CDV) opined pathologist Linda Munsen. CDV is a morbillivirus, a distant relative of human measles and the notorious rinderpest virus, which had decimated African wildebeest and water buffaloes a century earlier. In domestic dogs the RNA-containing virus causes seizures, nerve damage, and demyelination, an unraveling of the fatty insulating tissue that encases nerve fibrils. The CDV diagnosis was confirmed by Max Appel at Cornell, the world's expert on CDV, who used strain-specific CDV monoclonal antibodies to detect the virus proteins in lion brains. Margaret Carpenter recovered CDV gene sequences using the sensitive PCR-DNA photocopy technique in every sick lion she examined. By August 1994, 85% of the Serengeti lions tested positive for CDV and every sick and dying lion was infected with CDV.

At the time of the outbreak, CDV had been known to science for nearly two centuries. It was originally discovered by the vaccine microbiologist Edward Jenner in 1809. Typically CDV would cause disease in wild canine species (wolves, wild dogs, foxes), and also in raccoons, ferrets, skunks, and pandas. Experts believed CDV could infect cats, but would remain innocuous. Not this time!

Domestic animals are not allowed within the Serengeti, but Tanzania's indigenous Masai pastoral herdsmen were keeping over thirty thousand pet dogs on lands surrounding the park. British veterinarian Sarah Cleveland screened the Masai dogs' sera and found that about half of them had CDV antibodies. In one village on the eastern edge of the reserve the CDV prevalence in domestic dogs reached 75% at the peak of the lion outbreak. Domestic dogs and lions seldom clash physically, but spotted hyenas have bloody encounters with both. Sure enough, we were able to isolate CDV from the first seven Serengeti hyenas we tested. When Margaret Carpenter compared the CDV gene sequences from lions, hyenas, and the Masai dogs, they proved to be virtually indistinguishable. By contrast, the Serengeti CDV was quite distinct from CDV strains found in dogs elsewhere in the world. The Serengeti ecosystem had evolved its own "hot" strain and one with rather catholic species tastes. It easily jumped from dogs to hyenas to lions and took a heavy toll.

Craig Packer and Sarah Cleveland embarked on a campaign to vaccinate the Masai dogs with an effective attenuated CDV vaccine. They hoped to reduce the CDV reservoir in the Masai dogs, and thereby protect the vulnerable wildlife that faced annihilation. But, almost as abruptly as the outbreak started, the lions suddenly ceased dying in October 1994. Sick animals recovered, breeding resumed, and lions retook their place at the top of the fragile ecosystem.

Genetic differences among the outbred lions are most likely responsible for their survival. The lions demonstrate the insurance value of varied immune response within a genetically diverse population. We had documented an adaptive episode for CDV in freeze-frame accuracy. When the episode was over, all of us breathed a heavy sigh of relief.

It is not so often that an experienced scientific team can witness a brush with extinction up close. Without the Chandlers' video, Packer's surveillance, and Roelke's capture and veterinary team, we would have missed it entirely. But this time we caught it, watched it, studied it, and learned from it. Once again natural processes of gene interaction and parasite/host coevolution somehow forged an effective solution that allowed a natural population to survive an assault by a fatal microbe.

Although FIV turned out not to be responsible for the plague, it may nonetheless have played a supporting role. Did one or all of the three FIV strains (A, B, or C) weaken the immune system of its carriers and predispose them to the CDV mortality? Our latest studies have revealed that certain FIV-infected lions actually have relatively low CD4 T-lymphocyte cell populations, the hallmark of pre-onset AIDS in people living with HIV. Perhaps the double exposure to FIV and CDV exacerbated the effects of either alone. We are still trying to sort out these possibilities through more experimental data.

At this stage, the lessons from the lion plague come through loud and clear: Pay attention to the genetic variation of afflicted species. Try to witness and document natural disease defenses; they might offer strategies we would never uncover in a century of laboratory drug design. And watch the viruses and their genes. Pathogens are capable of amazing mutational changes. To survive and spread they must circumvent the immune defenses of their hosts. Infected animal populations must be flexible, evolving new defenses by selection of the most effective immune response artillery. Nature effectively promotes an arms race between pathogen and host. The war is often lost, but never won. Only the adapted fortunate live to join in the next encounter.

Medical or even basic research projects seldom proceed in a straight line; occasionally they turn full circle. The new biomedical technologies developed for human medicine have an unfulfilled potential to uncover the perils endangered species face. Without safe anesthesia, western blots, PCR, microsatellites, and virological monitors, we'd still be guessing what killed all those lions. Biomedical insight is critical for recovery plans designed to protect endangered species. Today's conservation management programs invariably include biomedical, genetic, and reproductive assessments. With better data and surveillance, we enable our conservators to deal with the real threats to species survival.

In return, wildlife survivors offer genetic solutions and adaptations that could be promising leads for human medical treatments. With incurable human diseases filling our hospitals and so many endangered species facing a rapid extinction vortex, uncovering these mechanisms would come none too soon. There are many reasons to understand and conserve wildlife species. The mutual benefit to animals and to ourselves in revealing genomic accommodations should not be underestimated.

Eight

The Wild Man of Borneo

SCIENTISTS DISPLAY A SURPRISINGLY COMMON TENDENCY for overstatement. I suggest it simply reflects our desire to be heard, understood, and believed. But there are a few well-known exceptions. The closing sentence in Watson and Crick's 1953 classic paper in which they first described the coiled double helix structure of DNA ended coyly, "It has not escaped our notice that the specific pairing we have postulated immediately suggests a possible copying mechanism for the genetic material." Witness the quiet drop of a scientific bombshell.

Another even more remarkable understatement of enormous scientific and cross-cultural import was a single sentence in Darwin's opus *On the Origin of Species by Means of Natural Selection* that suggested that his theory of evolution might apply to humankind. Although Darwin wrote not a word of his treatise about great apes (chimpanzees, gorillas, and orangutans), the world rapidly grasped the heretical notion that humankind has close kin among the monkeys.

People have always been fascinated by our likeness to the great apes. Many, however, have been fearful of apes. Novelists and filmmakers often depict them as horrific killers, as in Edgar Allan Poe's "The Murders in the Rue Morgue" or the Hollywood thrillers *King Kong* and *Congo*. That view has been tempered more recently thanks to the dedication of three talented behavioral researchers, Jane Goodall, Dian Fossey, and Birute Galdikas, who spent their entire careers in close observation of free-living primate communities.

Goodall's monitoring of chimpanzees in Gombe National Park in Tanzania, Fossey's gorilla studies in Virunga Karisoke reserve in Rwanda, and Galdikas's rehabilitation station for orangutans in Borneo each have produced accurate and detailed descriptions of the nuances of great ape societies. Revealed through articles in *National Geographic,* by television documentaries, and in detailed monographs, the world of the great and often gentle apes has become more visible in their parallels to the good and bad of our human spirit.

While the three "trimate" researchers were collecting their sociological details, molecular geneticists were debating a more elementary question: the evolutionary history and genetic relationships among the three apes and humans. Actually, the new discipline, called molecular evolution, driven by the principle of a molecular clock, cut its teeth on attempts to resolve human-ape species relationships. Most specialists could agree that the apes' earliest ancestors split off from the Old World monkeys (baboons, African green monkeys, mandrills, etc.) around thirty million years ago, and that the lesser apes (gibbons and siamangs, the brachiating apes) diverged from the great apes and humans approximately twenty million years ago. A rather good fossil record provided these datings, and the geneticists' DNA measurements fit with this time frame as well.

The more recent divergence of the modern great apes and humans proved more dicey. Experts by and large agree that the first great ape to split from the others was the Asian red ape or orangutan, which survives today in Indonesia and Malaysia on the islands of Borneo and Sumatra. That departure occurred around sixteen million years ago. The divergence of the African apes—gorilla, chimpanzee, and human—remained an unsolved mystery. Although man's closest relative was most certainly either the chimpanzee or the gorilla, it was not easy to decide which one was closer to humankind. Over a hundred scientific papers claiming to resolve the human-ape phylogeny, its evolutionary tree, appeared before a consensus finally emerged.

Man's closest relative appears to be the chimpanzee, which split from the lineage leading to humankind around five million years ago. Just before then (maybe six million years ago), but well after the orangutan divergence, the ancestors of the gorilla split away from the chimpanzee-human lineage. Further, there are two different living species of chimpanzee, the common chimp (*Pan troglodytes*) and the pygmy chimp or bonobo (*Pan paniscus*), which separated into their own distinct lineages around two million years ago.

These developments were of moderate interest to our laboratory group, mostly because we hoped some lessons from the primate research would help as we tackled the evolutionary hierarchy of our cats. Our new impassioned young graduate student, Dianne Janczewski, ratcheted up our interest in a hurry. Dianne had worked as the orangutan keeper at the National Zoo in Washington, and while there she learned firsthand of a major management fuss over raising these magnificent creatures in captivity.

Orangutans had been bred in zoological parks since the early 1970s, and by the late 1980s, there were 260 in American zoos. They were managed as a single population by a consortium of zoos, according to the orangutan Species Survival Plan (SSP). The orangutans bred well enough, but there was a subspecies purity issue at play. Of the captive animals, 30% originated from Borneo, 44% were from Sumatra, and 8% were intercrosses between the two or had an unknown geographical background.

Scientists believed the two Indonesian islands harbored different subspecies, *Pongo pygmaeus pygmaeus* in Borneo and *Pongo pygmaeus abelii* in Sumatra. However, the actual scientific criteria for subspecies recognition was murky at best. Delineations of morphological differences had been offered, but there were plenty of exceptions. Some described the Sumatrans as brighter orange with large, hairy male cheek pads, while Bornean orangutans were pigmented a darker red and had less mustache on their face. In truth, there was more difference between males and females on either island than there was between either subspecies, leaving experts feuding over whether the subspecies differences were important. One unequivocal genetic character did emerge from preliminary chromosome comparisons. All the Sumatran-born animals had one form of chromosome number 2, while the Bornean animals had a distinct, easily recognizable inverted segment of the same chromosome.

Zoo managers had been interbreeding the two subspecies for generations, but the SSP managers were worried that continued crossing between a genetically distinctive subspecies would destroy the evolved adaptation of the two island groups. Further, some predicted that mating between the two groups would introduce reproductive or developmental abnormalities that could cause congenital defects or even sterility.

So, was the intercrossing producing "mules" of two great ape subspecies? In February 1985, based on these concerns, the American Association of Zoological Parks and Aquariums (parent body of the orangutan SSP) imposed an indefinite moratorium on breeding between Bornean and Sumatran orangutans. Zookeepers were ordered to divorce lifelong marital couples because of the new dictum, although everyone acknowledged that little definitive evidence had been marshaled for or against the existence of two distinctive subspecies.

Dianne had fretted over this situation for long enough. No one paid much attention to her opinion in the debate because of her low status as a junior zookeeper, so she was determined to develop new defining data on the question, using the latest fine-tuned molecular genetic technologies. In the fall of 1986, she outlined her plan to collect DNA specimens from wild orangutans and to apply genetic insight to this perplexing issue, one with implications for captive and wild orangutan conservation management. She wanted to apprentice with our group to unlock the orangutan's misty past. I took her on.

Earlier that same year, across the continent in Washington State, a young, diminutive veterinary technician, Harmony Frazier, and the Seattle Zoo's crusty, muscular zoo veterinarian, Dr. William "Billy" Karesh, had a conversation very similar to mine with Dianne. Billy and Harmony would organize a DNA collection expedition to the tropical Southeast Asian home of the world's remaining wild orangutans. When Dianne and Billy discussed their parallel ideas at an orangutan SSP workshop, the seeds of a scientific partnership were born.

Billy would shortly relocate to head a Wildlife Veterinary Program at the Bronx Zoo, officially known as the Wildlife Conservation Society, one of the world's most influential conservation organizations. With their support, Billy packed trunks of veterinary gear—drugs, capture darts, tubes, diagnostics—and began an odyssey that most veterinarians only dream about. Documented in his book *Appointment at the Ends of the World*, he became a conservation advocate, a roving medical ambassador, and a collector of genetic goods. His first challenge would be procuring biological samples from the mysterious and solitary Asian great apes, a species Margaret Mead had once described as the "wild man of Borneo."

Billy Karesh and Harmony Frazier spent two years writing letters, cables, faxes, and petitions to Indonesian and Malaysian wildlife officials in their quest to get permission to collect orangutan DNA specimens. Their requests were filed, but no approvals ever came. Undaunted, they traveled to Jakarta, parked themselves in wildlife department offices, and pleaded to receive the requisite permits. Their patience and determination surely must have impressed Indonesian wildlife bureaucrats because they finally received the green light. They immediately hatched a plan for their first expedition to the tropical Gunung Palung National Park in Borneo. It was the beginning of a harrowing and treacherous adventure, one held together by their determination to carry home the precious tissue.

The orangutan (which means "old man of the forest" in the language of Borneo's indigenous Dayak peoples) is the only great ape to have evolved in Asia. Its historic range extended through Southeast Asia as far as India but was reduced to the islands of Borneo and Sumatra about ten thousand years ago. The periodic rise and fall of oceans over the subsequent millennia produced over seventeen thousand islands in the Indonesian archipelago and numerous isolated island species. Until very recently, the island of Borneo, which crosses the equator, was 90% tropical rain forest, the second largest in the world after the Amazon. Rainfall is perpetual in Borneo, in excess of 120 inches per year, enriching a massive jungle swamp ecosystem that includes a remarkable diversity of plant, insect, and animal species.

Orangutan numbers have diminished from over 100,000 a few centuries ago to between 20,000 and 30,000 today, subdivided into a few dozen populations within loosely connected habitats in Borneo and Sumatra. The census numbers continue to drop as a consequence of the destruction of habitat by forest harvest, wildcat gold-mining operations, and agrarian development. Protracted droughts in the early 1980s and then in 1997 led to the most devastating forest fires on the planet, leaving thousands of orangutans homeless. The orangutan species, comprising the Bornean and Sumatran subspecies, has been listed as Appendix I, the most extreme level of endangerment by CITES, since 1970.

In 1971, Birute Galdikas arrived in Borneo to begin a student research project focused on the most elusive of the great apes. Like Goodall and Fossey before her, she enjoyed sponsorship and mentoring from the famed paleontologist, Louis Leakey. Galdikas has meticulously described the orangutans' semisolitary lifestyle, social structure, and maternal nurturing, and the population fragility caused by increasing human exploitation of the land. With her husband, Rod Brindamour, she established Camp Leakey, a research site and "rehabilitation" center for rescued, confiscated, or otherwise orphaned orangutans. Galdikas managed Camp Leakey in the Tanjung Puting National Park of southwestern Kalimantan, Borneo, until the late 1990s. Her efforts to observe and interpret orangutan social behavior are described in stunning detail in her 1995 book, *Reflections of Eden: My Life with the Orangutans of Borneo*.

Galdikas took in orangutans orphaned by clear-cut forests, chemically assaulted gold-mining operations, and forest fires, as well as confiscated illegal pets. All told, she has rescued and returned over 200 young orangutans to the wild. Together with other rescue efforts in Borneo and Sumatra, close to 800 hapless orangutans have been rehabilitated and released over the past thirty years. Galdikas muses that when Joy Adamson released her orphan lion, Elsa, to the wild in the 1960s, the conservation community was charmed. And the fictional film *Free Willy* spun a tale of releasing a stranded killer whale to its home oceans to cheers from moviegoers worldwide. For orangutans; multiply the story of Elsa and Willy 800 times.

Orangutan rehabilitation programs, like most visible conservation programs, are not without vocal critics. Some worry that reintroduced orangutans will spread an exotic human disease picked up in the rehab centers to the wild population, but so far no evidence supports this fear. Others reason that releasing young orphans may disrupt a stable wild social system, but the carefully monitored populations show no signs of this stress. To the contrary, orangutans seem ideal for release because of their semi-isolated social structure. Unlike their African cousins, chimpanzees and gorillas, who live in interactive groups, there is not a great necessity for orangutans to be admitted to an existing troop. Perhaps all they need is a tall tree.

Birute Galdikas began to wonder if the separate populations on the two large islands or even isolated orangutan populations on Borneo had evolved away from one another enough to influence the success of her release programs. That is, if subspecies had become very distinct, then releasing orangutans in places distant from where they were born might be detrimental. The question was particularly thorny for confiscated orphans of uncertain origins. One could never really rely on a "busted" orangutan smuggler to disclose the true geographical origin of the animals.

When I met Birute for a long lunch in Los Angeles during one of her American speaking tours, she relayed her concerns. She asked that her orangutan family members be allowed to participate in the genetic study. Dianne and I were delighted to have Birute's orangutans aboard and we became more convinced that a population genetic assessment of the orangutans across Borneo and Sumatra would be important, no matter how it turned out.

Billy Karesh knew he couldn't anesthetize wild orangutans simply to collect a blood sample for DNA. They nested high in trees, and he did not have the luxury of a high-tech Florida-panther-style SWAT team or even a crash bag. So he designed a skin biopsy dart very similar to the one we had used on the humpbacks, but smaller for orangutans. The dart was four inches long, modified from a standard drug delivery veterinary dart. The point was a razor-edged cutting circle with two dental brush barbs in the center. Delivered by a regular veterinary CO_2 pistol to a large muscle, the dart cuts through the epidermis, snag-barbs the tissue punch, and falls off the animal. Other than a temporary sting, the animal is little distracted by the marauding. Billy had tested his biopsy darts first on leather jackets, then on several animals at the Seattle Zoo.

Dianne developed a method for freezing fresh skin biopsies in a fire-hydrant-sized liquid nitrogen tank. She equipped Billy with the tanks, life-support medicine, tissue collection supplies, and a detailed protocol. With supplies, drugs, darts, and permits in hand, Billy Karesh and Harmony Frazier set out to the Bornean jungles in search of orangutan DNA.

The first trip to Borneo for the biopsy team would later be described in Billy's log as their "voyage from hell." Harmony and Billy arrived at the sweltering and stench-filled Indonesian city of Teluk Melano during the Muslim holiday Ramadan. Their Indonesian counterpart and guide was a Muslim veterinarian from Jakarta. He had been fasting all day. As they ventured upriver in a rickety wooden canoe, he became delirious and collapsed from dehydration. When he came to, he admitted this was his first field trip.

The team's transport consisted of two wooden canoes, or sampans, powered by rusty eight-horsepower outboard engines prone to sheared pins and breakdowns. As they wended their way through the dense rain forest, they had to duck from diving giant fruit bats with four-foot wingspans and were serenaded at dusk by three-inch cicadas with decibel levels akin to a chain saw. Once they docked near the research station, a potpourri of leech species crawled onto their clothing as they hiked to the cabins at Gunung Palung National Park, taking away somewhat from the romance of the scenery. The leeches were not deadly, nor even particularly painful, but still distressing as they spilled blood meals on the researchers' shirts and trousers.

Sleep came quickly once they had dried and removed the leeches. The team set out early the next day to find the orangutans. After several days of lugging their capture gear through the lizard-, insect-,

and leech-infested swamp, they finally came upon their first orangutan. Billy fussed a lot, hoping to hit a broad muscle target while the
animal looked away. He carefully loaded the dart, took aim, and fired.
Success! Then came the hard part: finding the tissue-laden dart fallen
from the high branches to the swamp below.

After several days of hard work and discomfort, Billy, Harmony, the
Indonesian vet, and the guide began the return trip downriver,
relieved to have a cache of three orangutan skin biopsies. The final
calamity struck when the portable liquid nitrogen tank containing the
precious tissue samples was snagged out of the sampan by one of the
ubiquitous spiny rattan vines that adorned the riverbank. Harmony
was horrified and reflexively dove into the river, oblivious to potentially dangerous aquatic creatures, and rescued the tank before it
sank.

The samples were saved and shipped back to our laboratory for
analysis. Over the next several years on additional collection trips
across Borneo and Sumatra, Billy, Harmony, and Dianne netted over
fifty skin specimens from different populations including Birute
Galdikas's study site.

The molecular tools for assessing accumulated genetic differences
among populations or species had been improving steadily since the
early days when we screened blood enzymes in the African cheetahs.
DNA assessment methods including DNA sequencing were rou-
tinely being applied to evolutionary questions such as the human-
chimpanzee-gorilla trichotomy. Also, theoretical geneticists had
embraced the computer's awesome power to process voluminous data
and to reiterate tedious analytical routines for assessing population
diversity and genetic similarities among population groupings. The
computational programs were very precise and amazingly informative.

The orangutan's subspecies question was important and we were
determined to use every available approach to find the answer. As the
skin samples filtered in from Indonesia, Dianne thawed the tiny
pieces and planted the tissue in tiny plastic petri dishes bathing in
cellular nutrients. The skin fibroblast cells attached to the plastic sur-

face, divided, and produced a confluent cell layer limited in growth only by the size of the dish. Dianne tended to these cell aggregates like a seasoned gardener cultivating orchids. The growing cells were transferred to larger dishes until enough material was available for DNA extraction. Residual cells were refrozen to serve as an immortal genetic replica of the gentle apes in the Bornean forests.

The chromosomes of every tissue culture line were checked to confirm each orangutan's origin by identifying the telltale inversion on the second chromosome. We then moved to a molecular assessment of the genetic differences between the two island subspecies and between four Bornean populations that Billy, Harmony, and Birute had sampled.

Dianne first examined differences in allozymes, the gene products that we first used to assess variation in cheetahs. Next she examined 458 proteins using a radioactive protein stain applied to the electrophoretic gels of the living cell cultures. It took years for Dianne to collect and analyze her data, and to be certain of the conclusion, we added even newer genomic approaches. A determined new postdoc from Peking University in China, Lu Zhi, took the lead in screening the orangutans with four additional DNA methods, each aimed at comparing the orangutan populations. These were minisatellite DNA fingerprints, mitochondrial RFLP, and full DNA sequence analyses of two mitochondrial genes named 16s ribosomal RNA and cytochrome oxidase. Nearly ten years after we had hatched our plan to assess the status of orangutans, the genetic data were finally taking shape.

The results were dramatic both in consistency and in their ramifications. Irrespective of the genomic approach, the genes of Bornean and Sumatran orangutans were different—very different. We were recording clear divergence greater than any subspecies we had examined. The orangutan's genetic differences were on a species scale like lions versus tigers, horses versus donkeys, or green monkeys versus baboons. If our data were correct, it would mean that the two island orangutan groups were not subspecies at all. They were two different species. It was a rather remarkable finding. Could the wildlife science community have missed a true species of orangutan? It looked as if they had.

By then, we knew that genes and DNA sequences can evolve at different rates in different lineages of mammals. For example, rodents like mice and rats accumulate mutations in their genomes much more rapidly over a particular time interval than do primates or cats, in part because their generation time is shorter, providing more opportunity for mutational changes. So we needed to find a great ape species comparison to calibrate our assessment of the orangutans' large genetic distances. We chose two well-accepted great ape species, the common chimpanzee (*Pan troglodytes*), and the smaller, more upright pygmy chimpanzee or bonobo (*Pan paniscus*), which was restricted to a single rain forest in Zaire. In the same five gene systems assessed by Dianne and Lu Zhi, the genetic difference between the two chimpanzee species was about the same or in some cases somewhat less than that between Bornean and Sumatran orangutans.

We were also able to use Dianne and Lu Zhi's genetic distance data to estimate the time elapsed since the ancestors of Bornean and Sumatran orangutans had split from each other. The reference date we employed was a well-accepted date for the divergence of human ancestors from chimpanzees, 4.7 million years ago, recently estimated by molecular approaches in accordance with the age of great ape fossils. For each molecular method, we took the Bornean-Sumatran genetic distance, divided by the human-chimpanzee genetic difference for the same genes, and multiplied the ratio by 4.7 million years.

Our results were amazingly congruent. The estimated orangutan divergence date for Dianne's allozymes was 0.8 million years ago; for cell culture proteins, 1.0 million years; for Lu Zhi's mitochondrial DNA-RFLP, 1.5 million years; and for cytochrome oxide, 2.1 million years ago. An independent estimate by a Swedish lab estimated the Bornean-Sumatran divergence date as 3.5 million years ago based on complete mitochondrial DNA sequence (sixteen thousand nucleotide letters). These divergence date estimates may not look so precise, but all these tests put the split as one to three million years ago, plenty of time for a species-level differentiation to occur.

Paleontologists tell us that it takes between one and two million years for one species to evolve into a new species. No matter which genes we examined, the two orangutans were clearly of sufficient distinctiveness to be considered separate species. The data were unequivocal. Dianne, Lu Zhi, Billy, Harmony, and I published our findings together with our Indonesian collaborators. We recommended that both Bornean and Sumatran orangutans be considered as fully developed species.

But what about the separate populations on each island? Were there any population subdivisions that might influence Birute Galdikas's reintroduction locales? The genetic data were rather clear in showing little if any population-level genetic distinctiveness among five isolated populations in Borneo (Gunung Palung, Kutai on the west coast, Sabah in northern Borneo, Sarawak in northwest Borneo, and Birute's population in Tanjung Puting). We interpreted this similarity among the separate populations as reflecting rather recent gene flow or interbreeding between all the Bornean populations probably until interruption by human settlements within the last century.

Each orangutan population showed plenty of endemic genetic variation with allozymes, mitochondrial DNA, and DNA fingerprints compared to humans, chimpanzees, or gorillas. No recent bottleneck had reduced diversity in these populations as we had seen in cheetahs, Gir lions, or Florida panthers. These findings were encouraging for orangutan conservation. It was not too late to build a conservation plan to allow the two species to enjoy their genetic endowment.

Another implication was that conservationists need only to worry about the island of origin when reintroducing confiscated animals. Any Bornean individual could be released in any Bornean population without fear of genetic problems, because all the populations were by and large equivalent. This meant that release programs on each island could concentrate on nongenetic issues like habitat, carrying capacity, captivity, social behavior, infectious disease, and other ecological parameters.

Comparative studies of the hominoid apes, whether they target anatomy, physiology, behavior, or genetics, are beginning to shed light on our own human origins. The same molecular genetic tools we applied to orangutan population structure have revealed that modern humans trace their genetic legacy to African ancestors. As recently as 150,000 years ago, a group of African explorers ventured north to Eurasia and began peopling the world's continents. This may not have been the first migration of humans out of Africa; there is increasing evidence of at least two additional migrations before then by the progenitors of *Homo sapiens*. Detailing these events is beyond the scope of this chapter, but the parallels in genetic archeology are coming into focus. Neanderthals or other human "subspecies" are thought to have coexisted with and likely been extinguished by our direct ancestors. Until recently, we could only imagine the historic contact between human progenitors millennia ago. Now we have encrypted records, not only in the artifacts of archeological sites, but also passed down in our genes. Together the evidence is revealing some details of the trials and errors in the ancient human family tree.

An important social outcome of the human studies has been the ticklish genetic assessment of modern human ethnic groups. In the mid-1960s Harvard University population geneticist Richard Lewontin formulated a powerful approach to assessing the meaning of genetic diversity among modern human races. He asked what fraction of the enormous genetic diversity seen in human population represents race-specific differences and alternately what fraction is attributed to differences among individuals within a race. The answer to his question was determined by the computation of genetic distance between the three principal geographical human ethnic groups: African, European Caucasian, and East Asian. To get the answers, researchers have analyzed samplings of every genetic marker available, first blood groups (by Lewontin), then allozymes, mitochondrial DNA, microsatellites, and single nucleotide polymorphism (SNP) variants. In every case the results were the same and dramatic. Around 10% of human genetic variation depends on ethnic differences—which geographical race a person comes from. The other 90% of the variability is attributable to differences between individuals. This means that nearly all the differences in human populations lie at the individual level and only a small fraction represents true racial difference. Put another way, there could be ten times more genetic differences between you and your spouse than there are between Europeans and Africans. Lewontin and those that followed correctly interpret these results as an affirmation of genetic equivalence among the world's peoples, and by implication a definitive condemnation of a supposed genetic basis for racism, discrimination, and racist policy.

The new genetic technologies are in their infancy when it comes to revealing the secrets of our past. Our genomic code is a living hieroglyphics, more complex and intricate by orders of magnitude than the most elaborate spy code ever cracked by the National Security Agency. As a new breed of genetic interpreters cross biological disciplines in search of plausible scenarios for genetic patterns and disposition, they share the anticipation of learning from the missteps and adaptive solutions of ancestral species. Just as we must strive to avoid the holocausts and misdeeds of recorded history, can we now begin to uncover the missteps, prejudices, and pandemics of our very much older prehistory? I believe we *can*.

Nine

The Panda's Roots

OUR TREACHEROUS CLIMB LASTED SIX HOURS, SLIPPING AND sliding in the cold drizzle up the steep and muddy game path. Lugging our backpacks, in thermal undergarments and Gore-Tex hiking boots, we meandered through thick bamboo forest, over numerous precipices, by loggers' camps, past towering waterfalls. At each landing, the young and pretty Dr. Lu Zhi pointed her antenna toward the signal from a panda's radio collar. To reach this point, I had endured three days of plane travel to Xian and a ten-hour Land Rover excursion to Shanshuping Logging Camp. I was suffering from jet lag, Montezuma's revenge, and a mild hangover from rice wine the night before. Still, I was determined to relish an adventure few could even imagine: seeing a wild giant panda and her cub in their den, atop the Qin Ling Mountains in the Chinese mountain province of Shaanxi.

Lu Zhi was the first to hear the baby's call; it was faint, like a plaintive cry far off and audible only in the silence of the forest. At the pinnacle, Xiao Jian, a young graduate student, led us to a massive boulder overhanging the steep cliff below. By now the baby's cries had grown uncomfortably shrill. Lu Zhi agilely descended around the boulder to approach the panda's den below. Professor Pan Wenshi, Lu Zhi's mentor from Peking University, helped me around so I could view the tiny eight-inch-long creature howling with the pain of separation from her mother.

The baby giant panda was unmistakable with its alluring eyespots and Mickey Mouse ears, lovable and fragile. They had named her Gui Ye.

Xiao Jian explained that Gui Ye's mother, Mo Mo, was out foraging for food, a common activity documented in wild pandas by Pan Wenshi and Lu Zhi during their thirteen-year-long study of giant pandas in Qin Ling.

I could not take my eyes off the fragile cub, active, loud, and adorable. Professor Pan asked if I would like to hold the cub briefly; he assured me it would not hurt the baby. I carefully reached for the tiny black-and-white infant and drew her slowly with both gloved hands to my warm parka. Her squawking subsided as she closed her eyelids and rested her head in the comfort of my scarf. The others smiled quietly because they knew what I was feeling.

As I held Gui Ye to my chest, it occurred to me that this is what species conservation is really about. That moment is as sharp in my memory as if it happened yesterday. After the cub was replaced carefully in her den, I gazed wistfully across the remote alpine valley until my nirvana was interrupted by a sudden screech from Gui Ye. Lu Zhi had plucked a single hair from her backside for a genetic sample.

In retrospect, I guess it was inevitable to be drawn into giant panda biology and conservation. Pandas are the unchallenged flagship species for animal conservation campaigns. No other endangered species has attracted as much attention, study, dispute, or passion as the shy, alluring, black-and-white "cat-bear," as the animal is called even today in China. Pandas are the superstars of zoos and conservation initiatives. Perhaps because of their tremendous popularity, they have also attracted the very best—and the very worst—of conservation saviors to help them prosper and avoid extinction. So many TV specials, books, and articles on the politics, science, value, and history of this species have appeared that I am hesitant to add yet another chapter to the library of giant panda narration. Yet, the giant panda's genetic story is a fascinating biopolitical journey, one from which we can learn by the missteps as well as the successes.

The most famous giant pandas in my part of the world were Ling Ling and Hsing Hsing. The pair arrived at Washington's National Zoo on April 16, 1972, a gift from Premier Chou En-lai and the People's

Republic of China to the people of the United States to commemorate the opening of China to Western diplomacy by President Richard Nixon. The Washington pair received enormous public attention and affection. Their first day on exhibit attracted seventy-five thousand people. Over a million visitors came to see the playful youngsters before the end of 1972.

Throughout the 1970s, every American hoped for a baby panda, but it was not to be. Giant pandas breed seasonally, coming into estrus (a period when females are receptive to a male's affection) only once a year, usually in the springtime. Ling Ling and Hsing Hsing were put together every season beginning in 1974, Ling Ling's first estrus. But Hsing Hsing had trouble with his part, often mounting the wrong end, ejaculating excitedly into Ling Ling's ear or in other comical but ineffective locales. In the flurry of media surveillance, Hsing Hsing was labeled inept, a bumbler, a wimp.

By 1981, National Zoo curators and panda reproduction experts were seriously concerned. They airlifted Chia Chia, a virile, studly giant panda from the London Zoo who had fathered a cub with a female in Madrid, to provide some proven experience to the breeding challenge. Chia Chia's encounter with Ling Ling was a disaster. He thrashed the American female senseless, sending her into shock. To stop the assault the zoo staffers fired rifles in the air and sprayed fire hoses at the pair. There was no mating, only marauding, and Chia Chia was dispatched back to the United Kingdom in disgrace. "The bastard mauled her," gasped National Zoo director Ted Reed. *Washington Post* columnist Mary McGrory dubbed Chia Chia as the panda equivalent of Stanley Kowalski—the brutish character from the memorable Tennessee Williams stage play *A Streetcar Named Desire.*

Shaken by the brawl, zoo managers decided to try artificial insemination, a procedure by means of which Chinese zoos had achieved success for pandas. So when Ling Ling came into heat in 1983, London Zoo veterinarian Dr. John Knight was poised to collect semen from Chia Chia by electroejaculation. Knight carried the fresh semen in his breast pocket aboard a British Air flight to Washington the next day. Before Chia Chia's sperm arrived at Dulles Airport, National Zoo's reproductive expert Devra Kleiman decided to give Hsing Hsing one last chance for a natural mating. Under her watchful eye, Hsing Hsing accomplished what resembled a proper mating mount and Kleiman heard an ecstatic panda squeal, a behaviorist's signal of copulation climax. Just the same, later that day, Ling Ling was anesthetized and inseminated artificially with sperm just delivered from London's Chia Chia.

At the press conference describing these events shortly thereafter, a reporter asked Bob Hoage, the zoo's public relations spokesman, how they would know who the father was if Ling Ling was indeed pregnant. Hoage answered, "We have a longtime and close collaboration with the National Cancer Institute's geneticist, Dr. Stephen O'Brien. I am sure he will be able to sort this out." Ling Ling did get pregnant and I found myself tossed into the panda's conservation society for the first time.

Ling Ling gave birth to the first baby giant panda born in the United States on July 21, 1983, a tiny hairless mouse-looking infant of 134 grams. Sadly, within three hours of birth, the cub died of a *Pseudomonas* bacterial infection acquired in the birth canal.

The baby's autopsy was a grim event. I quietly collected a dime-sized skin tissue biopsy that we used to establish a living culture for genetic typing. We already had skin cell fibroblast lines from Hsing Hsing and Ling Ling. A week later, a skin snip from London's Chia Chia immersed in tissue culture medium arrived by plane. We tested twenty-nine allozyme genes in the panda's cells, selected because they were genetically variable in other mammals. Unfortunately, none of these genetic markers showed any variation among the four pandas. We needed genetic differences to establish the paternity of the baby, so we decided to use a larger group of proteins from the skin cell lines that were assayed by a protein stain after separation by a two-dimensional gel electrophoresis procedure. Of the three hundred proteins we examined, six showed genetic variation, and these proved that Hsing Hsing, not Chia Chia, was the baby's father. The natural mating witnessed by behaviorist Devra Kleiman had been successful.

With knowledge of Hsing Hsing's virility, the zoo attempted natural matings each year thereafter. Three subsequent pregnancies occurred, but like the first, the infants never survived, usually due to a massive bacterial infection. In 1992, Ling Ling passed away at age twenty-two; Hsing Hsing died in 1999.

Involvement with the world's first giant panda paternity assessment proved to be educational for me. As I lingered in the National Zoo's giant panda exhibit awaiting tissue samples, I absorbed the graphic displays. Elaborate wall decorations described the rich history of giant panda conservation, but also a century of cantankerous scientific debate. Taxonomists and panda biologists simply could not agree whether the giant panda should be classified in the bear family, Ursidae, or in the raccoon family, Procyonidae. As a compromise, some placed it in a separate family of its own.

The colorful saga of the panda's classification goes back to the first description of the giant panda by a Westerner, Père Armand David, a nineteenth-century Basque missionary. An amateur naturalist, David relished in documenting numerous mammal species endemic to China. In 1869, he sent a description of the mysterious alpine creature, which he called *Ursus melanoleucus* (black-and-white bear), to his scientific mentor Alphonse Milne-Edwards, director of the Paris Museum of Natural History. The following year, David brought back pelts and skeletal material for the museum. Milne-Edwards examined the animal bones and teeth and concluded that they more closely resembled those of the giant panda's diminutive Chinese cousin, the red or lesser panda, than those of other bears. Since the red panda had been placed in the raccoon family, Procyonidae, with almost no argument, Milne-Edwards proclaimed that the giant panda should also be classified as a giant raccoon, Procyonidae, along with the red panda. This exchange began a century-long argument that dominated zoological discussions in lively and provocative ways.

Although the giant panda certainly looks like a bear, it has some very unusual features related to its isolated alpine lifestyle and a highly specialized diet, primarily consisting of bamboo shoots, stems, and leaves. The giant panda's skull and jaws are massive; its powerful jaw muscles and broad teeth are suited to crushing and grinding its vegetarian diet. Giant pandas have a moderately developed sixth digit—the panda's thumb—adapted for grasping bamboo shoots. Red pandas live on bamboo as well, even gripping the shoots in the same way, but they lack the extra thumb.

The giant panda has several additional features that bears do not. Their male genitalia are tiny and backward pointing, similar to raccoons. Most bear species of temperate and Arctic regions exhibit a type of winter hibernation, whereas the panda does not. This might be because pandas are unable to store enough energy from bamboo, a rather inefficient energy source. Bears roar or growl; the giant panda bleats rather like a sheep or goat. The bleat actually has an equivalent in the twitters and chatters of red panda and other procyonids, but nothing similar is heard from bears.

Over the past 130 years well over fifty learned treatises, each claiming to resolve the panda's origin and its correct taxonomy, have appeared in the scientific literature. The experts split their decision into three camps, each dramatically opposed, but nonetheless dogmatic in their absolute certainty. One group declared the giant panda a specialized bear, Ursidae. A second considered it an unusual raccoon, Procyonidae. The middle-of-the-road group placed giant panda and red panda together in a separate family (Ailuridae or Ailuropodidae), emphasizing their biological distinctiveness apart from both bears and raccoons. As I waited for my skin specimen in the house of Ling Ling and Hsing Hsing, I realized that the only agreement about the panda's evolutionary roots was that there was no prevailing consensus.

So I began to read up on the panda classification opinions. And I have to say, I was impressed not only by the depth of the scientific debate, but also by the eloquent discourse of the warring factions about the meaning of the dispute. Edwin Colbert, who favored the Procyonidae school, opined in 1938:

> So the quest has stood for many years with the bear proponents
> and the raccoon adherents and the middle-of-the-road group
> advancing their several arguments with the clearest of logic,

while in the meantime the giant panda lives serenely in the mountains of Szechuan with never a thought about the zoological controversies he is causing by just being himself.

Stephen Jay Gould, affirming the bear school in 1986, remarked:

We revel in our persistent failure to resolve certain key arguments, because all the fun is in the fighting.

George Schaller, favoring a separate family, in 1993 wrote:

The panda is a panda. . . . In some ways the long-running issue is trivial, an illustration of scientific discomfort with uncertainty and a penchant for putting everything tidily in place. . . . However, the controversy poses a fundamental scientific problem— what features are important and significant in classification?

John Seidensticker in 2001 mused:

Scientists' efforts to understand the origins of the giant panda . . . have become legendary, as shrouded in misunderstanding as the animal itself. . . . It is fascinating to trace how different lines of reasoning and evidence unfolded as . . . newer, more sophisticated techniques emerge. . . . This is science at work; never static, always questioning, trying yet one more time for clarification.

Never has a taxonomic dispute been debated with quite such intensity, such elegant rhetoric, and such a rich store of morphological, behavioral, and anatomical descriptions. Consider that in 1964 D. Dwight Davis published a magnificent opus on the anatomy of Chicago's Brookfield Zoo panda Su Lin, who had died in 1938. His three-hundred-page analysis of fifty organ systems, replete with exquisite detail in text and illustration, was described by Stephen Jay Gould as "our century's greatest work of comparative anatomy." Davis's taxonomic conclusions were resounding: "The giant panda is a bear and very few genetic mechanisms—perhaps no more than half a dozen—were involved in the primary adaptive shift from *Ursus* (bear) to Ailuropoda (panda)."

Davis's view was quickly accepted by many, but alas not by all. His critics suggested that his analysis, however thorough, did not follow standard taxonomic procedures and simply detailed a collection of subjective judgments around his observations. On the first page of his monograph Davis admits in a footnote that he became convinced early on that the giant panda was a bear, so he simply would assume this throughout his text. He made no attempt to present a comprehensive comparative analysis of giant pandas to red pandas, raccoons, and bears because, in his words, "this became so difficult I gave up." Davis's work was masterful and persuasive, but more descriptive than analytical. His Achilles' heel was his presumption that his colleagues would embrace his interpretations simply by the breadth of his detail. Most did, but many did not.

I could not resist this century-old debate, unsettled and looking for a new approach. Molecular genetics would surely weigh in and I had all the necessary biological materials in the cell lines: an immortal source of DNA, proteins, and genes from giant pandas, red pandas, and bears.

In the last few decades the new molecular evolution approach for solving phylogenetic history and taxonomic relationships had matured. The tactic compares genetic similarities and differences, particularly from homologous gene sequences and protein gene products (amino acid sequences) of several related species. The patterns of gene sequence divergence are used to reveal the ancient evolutionary splits that led to living species groups. To construct a species' evolutionary history, its phylogenetic tree, we return once again to the molecular clock hypothesis, which holds that the longer two species have been separated, the greater the amount of mutational divergence. Working backward with a matrix of genetic differences between species, sophisticated mathematical methods are used to reconstruct the evolutionary tree that best accommodates the data. In fact, there are now several theoretical methods (named, distance based, Basean, maximum parsimony, and maximum likelihood) that use computers to build phylogenetic trees from molecular data.

When the same results are achieved with each analytical method, particularly with several different DNA stretches, one can be reasonably confident that the relationships they specify reflect the true evolutionary history.

The molecular clock hypothesis has been met with a certain amount of controversy. Critics point out that the accumulations of mutations do not always occur precisely in proportion with time, causing the molecular clock to produce evolutionary trees with speed-ups and slow-downs in the limb lengths. These criticisms are troubling, but they do not invalidate the approach for determining relationships. Instead they emphasize the need to replicate tentative resolutions using different genes or DNA stretches. After forty years of application, improvement, and validation, the molecular clock methods have stood the test of time. They work remarkably well.

Since immortal living cell cultures from the giant pandas were growing in my laboratory, we applied six different molecular genetic approaches to the question of the giant panda's origin. Each method had distinct strengths and weaknesses, but we were confident that the six complementary methods would, when gathered together, give us a strong answer.

Three of the genetic tools—allozyme genetic distance, 2DE genetic distance (the same methods used for Ling Ling's paternity), and immunological distances—assess amino acid sequence changes in a group of proteins that have accumulated between the different species. The differences in amino acid sequence of the same protein (e.g., hemoglobin or albumin) in two different species is a consequence of mutational divergence in the homologous genes that encode the proteins. So, in effect, we are tracking evolutionary differences in genes indirectly by quantifying the extent of sequence differences.

Two methods, DNA hybridization and mitochondrial DNA sequence analysis, employ DNA sequence differences in homologous genes. These procedures directly assess gene divergence as a surrogate for evolutionary time.

The last method examines differences in the organization of chromosomes compared among different species. Over long periods of evolutionary time, ancestral chromosome arrangements are occasionally rearranged by spontaneous breakage and rejoining—a process called "translocation." Closely related species would have fewer chromosome translocation breaks, while the long period of separation for distantly related species would result in more translocation exchanges. By examining the appearance of chromosomes and identifying the regions that have the same homologous genes in two separate species, we can count how many chromosome translocations have occurred and identify where they happened in the chromosomes. Using the pattern and frequency of these chromosome translocations between several species, we deduced the chromosome rearrangements that occurred between bear and panda species. Then we applied an evolutionary principle called "maximum parsimony" to the steps in chromosome exchange, which would best explain the chromosome segment movements that led to the modern species chromosome organization.

The six different methods each gave the same answer to the bear-panda-raccoon evolutionary puzzle. Together they offered a graphic description of the evolutionary history of the pandas. And Père David got it right. The giant panda was indeed a bear; its genes consistently stemmed from the branch of an evolutionary tree leading to the bears, while the red panda's genes stemmed from the raccoon-procyonid branch. The new data revealed with some level of clarity how and when each species did evolve. Here is the story that the molecules told us about how the bears and pandas came to be.

Approximately thirty million years ago during the geological epoch called the Oligocene, the ancestors of today's eight bear species (brown bear, black bear, Asiatic black bear, sun bear, polar bear, sloth bear, South American spectacled bear, and giant panda) and procyonids (nineteen species, including raccoons, coatis, kinkajous, olingo, and red pandas) split apart to evolve separately. Less than ten million years later an ancestor of the red panda split off from the other procyonoid species to evolve in Asia while the rest of the raccoon family of species would evolve in the Americas.

Many years after the bear progenitors had split from the procyonid species—we estimate it at between eighteen and twenty-five million years ago—the giant panda's ancestors split away from the other bears. The giant panda gradually evolved its special adaptations for its alpine habitat. On the bear lineage now separate from giant pandas, the next fork led to the spectacled bear of South America (*Tremarctos ornatus*) on one side and a group of the other six bear species on the other. That evolutionary splitting event took place twelve to fifteen million years ago. The six modern ursine bears—brown bears (*Ursus arctos*), sun bears (*Ursus malayanus*), sloth bears (*Ursus ursinus*), Asiatic black bears (*Ursus thibetanus*), American black bears (*Ursus americanus*), and polar bears (*Ursus maritimus*)—subsequently diverged from one another about six million years ago. The breakup of the six ursine bear species was almost contemporaneous, making it difficult, even using powerful molecular techniques, to resolve exactly how they are related among themselves.

Although genetics can produce conclusive relative relationships, the dates we estimated for these splits were made possible by paleontologists who study fossil bears. Fossil specimens not only confirm the molecular tree but, because of their geological dating, supply a time scale. For example, the *Agriarctos* fossil is generally believed to be an early ancestor of the giant panda line, and that species occurred during the Miocene epoch about fifteen million years ago. The common ancestor or "missing link" for the nonpanda ursids is *Ursavus*, dated at twenty to eighteen million years old. *Plionarctos* is a suspected ancestor of spectacled bears from the late Miocene, some six million years ago, while *Ursus minimus* (about five million years old) is thought to be a primitive ursine bear ancestor.

I was deeply impressed by the clear resolution the molecular genetics had produced and by the near complete agreement among six quite different approaches. Further, the scheme fit very well with the fossil specimens and their time of occurrence.

Was it presumptuous of me to think we had solved a 120-year-old evolutionary puzzle by what some observers would describe as "slick and ultramodern techniques"? Perhaps, but I believe there is an important reason the molecular approach resolved the panda's roots when other more traditional analyses had stumbled, a reason that also explains why molecular genetic approaches have shed new light on other controversial evolutionary conundrums.

When a comparative biologist examines several species in a group, he or she finds some homologous traits, which provide evidence for close association in evolutionary time, and some analogous traits, which confound the process.

Homologous traits are characters inherited from a common ancestor that when examined in detail usually show multiple complex and detailed similarities. For example, all mammals have body hair, are nourished by mother's mammary glands, and have four-chambered hearts. These are homologous characters passed down from primitive mammalian ancestors. Members of the cat family, Felidae, have retractable claws, toes on their hind feet, and pupils that contract. The details of the anatomy and physiology of these traits are the same in different cat species, evidence that they also inherited them from a common ancestor.

Analogous characters look similar on the surface, but their developmental basis is different because they evolved at separate times on independent evolutionary lineages. Wings on insects, on birds, and on bats are analogous: same function—flap to fly—but independent origins.

Evolutionary relationships are quantified by counting the number of homologous trait similarities among a group of species. The greater the number of shared homologous traits, the more closely the species are related. The trouble with morphological characters like those described in anatomy and fossils comes when one tries to decide whether a particular character is homologous or analogous. If we mistake analogous similarities for homology, we will misinterpret a species' evolutionary past. The giant panda and red panda's shared method of grasping bamboo is a good example.

Molecular data make distinguishing homology from analogy much easier. One simply aligns the sequences and identifies homologous sites by their position in the linear script of DNA letters or amino acids. The more homologous characters examined, the more likely the phylogenetic result. The giant panda analyses looked at some

500,000 DNA letters plus the chromosome translocations that occurred over the fifty million years of bear-procyonid evolution. Molecular genetic traits are virtually unlimited because a mammal species' genomes contain over three billion DNA letters, each evolving according to a tempo set by an inborn molecular clock. So if we get a confusing result, we simply sample more genes until we arrive at a more convincing solution.

Also, the majority of DNA nucleotide letter variation is genetic "noise," not subject to strong natural selection. The influence of such variation on an individual's survival or fitness is considered neutral. This means that each difference can be given about the same weight as all other single changes. By contrast, visible morphological or anatomic distinctions, adaptations like the panda's thumb or its grinding herbivorous teeth, are not easily weighted or quantified with respect to their role in natural selection. There is no way to measure how many genetic changes contribute to an adaptive divergence. For example, Kodiak bears are, on average, twice the size of grizzly bears. Is the size difference a major adaptation that might suggest species divergence? (No, both are subspecies of brown bear.) This uncertainty can seriously confound taxonomic conclusions based solely on morphological characters.

We published our molecular solution to the giant panda's origins in *Nature* in September 1985. The magazine featured adorable photographs of the giant and red panda on its cover. In 1987, I wrote a popular version of the article for *Scientific American* because I wanted the authors of basic biology textbooks, who borrow heavily from this excellent magazine, to appreciate the details of our findings. On the *Scientific American* cover was a brilliant painting of a giant panda's massive and lovable head.

Perhaps due to the visibility of giant pandas and to the endurance of the debate, our announcement sparked strong reactions. Much comment was laudatory and many welcomed the large, comprehensive, and apparently definitive new data set. Some classical taxonomists were perturbed by all the hoopla, since as far as they were concerned, the giant panda's phylogenetic position was settled by Davis's monograph in the 1960s. I was, of course, pleased that even our critics agreed in principle with our conclusions. To counter their irritation over the new publicity I simply pointed to a dozen-odd reports published after Davis's that still disagreed with the bear school.

One gathering of giant panda experts from across the globe in 1991 was particularly memorable. The conference, held in the 4H Conference Center in Front Royal, Virginia, welcomed a large delegation from China and we all wore headphones to benefit from simultaneous translations. In the session dedicated to giant panda taxonomy, I outlined our molecular genetic results to polite but not-so-energetic applause. The next speaker was Professor Hu Jinchu, a Chinese field biologist who had worked in the Wolong reserve for years tracking giant pandas. He was coauthor with George Schaller of the seminal 1985 book, *The Giant Pandas of Wolong*.

Hu Jinchu had decided, based on his extensive behavioral and morphological observations, that giant pandas and red pandas are indisputably similar to each other and should merit classification in a separate carnivore family, Ailuropodidae, distinct from ursids or procyonids. He did not mention our contradictory genetic results. Professor Hu presented, among others, one remarkable image in support of his claim: a photograph of two fecal droppings—one from a giant panda, an oblong torpedo shape, and another from a red panda, a similar shape about one-quarter of the size. His implication was that their similar appearance supported the recency of red and giant panda's common ancestry. I knew I had my work cut out for me.

After the platform presentations, the speakers were brought to the front facing the audience for a discussion session. Professor Hu sat at one end of the speakers' table; I was at the other end, dreading the coming exchange. Hu began reiterating his position based on what I considered tentative if not weak data. Before I could respond, another speaker, Pan Wenshi from Peking University, tore into him in a faster, louder Chinese cadence. He characterized our molecular genetic results as comprehensive, definitive, and compelling. The scientific community should recognize the quality and unequivocal implications of the new results and stop fighting over superficial morphological similarities. Molecules don't lie; the issue should be closed. I did not have to say a word; Professor Pan had turned the tide. The debate had ceased. I just smiled inwardly.

Soon thereafter I dashed to shake Professor Pan's hand to thank him. He was accompanied by his graduate student, Lu Zhi, who deftly translated my English to his Chinese and Pan's Chinese to English. We eagerly chatted late into the night. I learned that Professor Pan had been working with giant pandas for twenty years, starting in Wolong with Hu and Schaller (Pan also was an author on their book), but was self-trained in field biology. His and Lu Zhi's original background was in biochemistry.

From Wolong, Pan Wenshi went on to lead a giant panda field study in the Qin Ling Mountains, home to 230 of the world's 1,100 remaining giant pandas. He had tracked pandas through the frigid alpine habitat and methodically monitored life history, behavior, mating, foraging, and playing of a group of giant pandas continually for sixteen years. Pan was passionately dedicated to giant panda conservation and convinced that the only way to cut through the fog of hyperbole, mistrust, and misinformation that permeated giant panda biopolitics was to get the science right and to communicate the results honestly, free of political agenda.

Pan was not shy. He pulled no punches in his unabashed support for high-quality work or his disdain for superfluous rhetoric. In the coming years Pan and I grew to become close friends. We collaborated, we shared data, and we often laughed together. In the future we would travel together across China, all the time discussing his findings, our opinions and preferences, and conservation science. As our first evening wore on I gradually became more aware of Pan's hidden strength, his diminutive, self-effacing student interpreter, the twenty-five-year-old Ms. Lu Zhi.

Lu Zhi was the prize-honored daughter of university professors from Lanzhou, three hundred miles northwest of Beijing. Her parents had been forcibly separated during the tumultuous Cultural Revolution in China (1966–1976), so Lu Zhi spent many formative years watched by her grandmother. She had enjoyed little time with her parents before she migrated, at age sixteen, to Peking University, the most prestigious and competitive academic enclave in China. She shone at Beida (as Peking University is nicknamed) and in her third year discovered the giant panda field project sponsored by the charismatic Professor Pan Wenshi. She knew right away that she needed to be a part of this study for the pandas, for China, and for her own fulfillment.

In 1985, Pan and Lu established a base camp at Shanshuping, high in the alpine bamboo forests of the Qin Ling Mountains. They tracked pandas on foot, recording every movement. In time, conservation organizations offered financial support for radiotelemetry equipment and transmitting collars. They placed radio collars on seventeen pandas and recorded the movements of eighty pandas, documenting eleven births and four natural mortalities by 1992. Their study would come to be the most extensive natural history of the giant panda ever recorded.

In Lu Zhi's spare time in the mountains she self-trained into a world-class wildlife photographer. Her gorgeous wild giant panda photographs were featured in two issues of *National Geographic* in 1993 and 1995.

Pan and Lu explained to me that first night in Front Royal that Lu Zhi was ready for a postdoctoral fellowship overseas. Afterward, she would return to China's conservation effort with international knowledge and experience. Both of them were certain that the new genetic technologies would be an important component of any conservation effort. Would I be willing to sponsor Lu Zhi in my laboratory for a few years to train her in molecular population genetics? Could she take a close look at the genetic structure of free-ranging giant panda populations? Were the pandas suffering from a population crash and genetic impoverishment like the cheetah or Florida panther?

It took me a nanosecond to agree. What a chance for us all. Train the next generation of conservation practitioners in the intricacies of molecular genetics? Get a chance to examine scores of blood and tissue samples from the elusive giant pandas? Maybe see a bit of China Westerners have rarely seen? I was thrilled. Lu Zhi arrived at our lab in the fall of 1992.

We taught Lu Zhi molecular genetics, starting with the aforementioned orangutan species question for practice; then she moved on to giant pandas. Her training in our laboratory brought new meaning to Richard Rodgers's memorable line: "If you become a teacher by your students you'll be taught."

Across the years, I would make several trips to China with Pan and Lu. We traveled to their study site, to panda reserves, to panda meetings, and to their homes. They opened their experiences, their past, and their hearts to me, a spellbound student of conservation's fits and starts, all against the backdrop of the tremendous social changes going on in China.

The giant panda conservation story was troubling, but complex. The remaining giant pandas were dwindling. Censuses in the 1970s and 1980s estimated that eleven to twelve hundred pandas survived in the wild, restricted to six alpine-forested regions on the eastern edge of the Tibetan plateau in western China. The surviving populations were subdivided to about twenty-four small populations, separated by mountain ranges, rivers, roads, forest clearings, and human settlements. In the past twenty-five years the range occupied by giant pandas has been cut in half. Nearly half of the residual populations number fewer than twenty individuals. Such small populations are at significant risk for extinction due to chance events such as the freakish death of a dominant breeding male, poaching, an infectious disease outbreak, or a climatic catastrophe. Over recent decades, about one small population had blinked out every year.

In the mid-1980s World Wildlife Fund International (WWF) formed a partnership with the Chinese Ministry of Forestry to develop the National Conservation Management Plan for the Giant Panda and Its Habitat. The management plan was composed of a power-packed group of field biologists, conservationists, and scientists, both Chinese and Western, coordinated by British conservation biologist John MacKinnon. I had a small role in drafting this plan during a three-week visit to Wolong Giant Panda Breeding Center near Chengdu in 1985. I managed to get a few paragraphs on the potential genetic problems due to inbreeding tucked into the plan. But the principal threats to giant pandas were not genetic; they were people, people, and more people.

The population of China grew from 450 million at the beginning of the communist revolution to 1 billion by 1982, stimulated by Mao Zedong's encouragement to increase the population. Today there are 1.26 billion people in Mainland China, roughly 1 million people for every single giant panda. Until recently, thirteen giant panda reserves established in the 1980s protected about 60% of wild pandas, yet thousands of people also lived in the reserves. A major threat to pandas was habitat elimination caused by a flourishing lumber industry within and outside the panda reserves. Poaching has also been an important factor: Panda pelts can bring prices of over ten thousand dollars in the underground wildlife trade of Asian capitals, a hundred times the annual income of a rural Chinese farmer. In Qin Ling, Pan and Lu saw six pandas lost to poachers over a ten-year interval.

The tragic government-implemented massacre of a few hundred protesting students and workers at Tiananmen Square in downtown Beijing in June 1989 had a chilling effect on all Sino-Western dialogues, including those involving giant panda conservation. Yet, I came to learn over time that many Chinese ministers, people, and children loved the animals deeply and were determined to protect them irrespective of political turbulence. Finally in 1992, the Chinese government endorsed the Chinese-WWF giant panda conservation and management plan and pledged to provide twelve million dollars of the projected eighty-million-dollar cost. Pan, Lu Zhi, and I thought that the balance might be raised by lucrative giant panda breeding pair loans. By the early 1990s captive giant panda pairs from Chinese zoos and panda reserves sent to Western zoos on "breeding loans" were netting a hefty ten-million-dollar donation each to Chinese giant panda conservation programs. We began to lobby publicly and privately for the loan fees to go toward the WWF-Ministry of Forestry conservation program.

At a 1993 Giant Panda Conservation Assembly in Chengdu, Pan and Lu Zhi gathered twenty-nine top scientists' signatures for a petition made to Chinese Premier Li Peng to suspend all timbering in designated panda reserves. Within a month, a courageous positive response came from Vice Premier Zhu Rongi. Logging was to be suspended indefinitely in Qin Ling and three million dollars was budgeted to relocate twenty-three hundred lumber workers and their families living there. Shortly thereafter, following the WWF-Ministry of Forestry master plan recommendations, the giant panda's protected area was expanded from thirteen to thirty-three national reserves. Timber harvest was prohibited inside all the reserves and WWF-trained antipoaching rangers were recruited to patrol panda habitat.

Lu Zhi's genetic study of wild pandas revealed that three mountain range populations (Qionglai, Minshan, and Qin Ling) all retained appreciable levels of genetic diversity. This was good news because it meant the populations had become isolated and small only very recently, within the last centuries. So far they were not showing signs of severe inbreeding like we had seen in cheetahs and Florida panthers. If their habitat, lifestyle, and solitude could be preserved, their genetic endowment looked very promising.

At the end of 1995, Lu Zhi returned to China, where she became the leading conservation coordinator for WWF's giant panda program. She labored through vast, isolated panda reserves, befriending the rangers, training them in field techniques and wildlife management and encouraging them in the nobility of their lonely important work. In 1999 the Chinese government named her as one of the top ten young Chinese citizens, recognizing her strong, positive influence on China's future. Pan and Lu Zhi, as much as anybody, embody the very best of their two generations' conservation champions.

It was disputed paternity and contentious zoological debate that drew me to giant pandas and their curious natural history. But our molecular research carried us beyond an esoteric scholarly pursuit to solve a long-standing taxonomic issue. Our results confirmed and extended the brilliant insight of D. Dwight Davis and demonstrated better than ever before the power and precision of the new molecular techniques. Our odyssey has made its way into biology textbooks and I believe laid important groundwork for the many far more sophisticated molecular phylogenies to follow.

The endless and often tedious battle between molecules and morphology has been derailed. As Harvard evolutionary biologist Ernst Mayr quipped to me a few years back, and I paraphrase: "You molecular guys go ahead and supply us with the evolutionary trees. Then we can get on with the really interesting challenge of interpreting the morphological changes by which adaptation has molded living and extinct species."

Touché. How right he was. And therein lies a prescient instruction for today's students of evolution. Develop a familiarity and critical expertise in each of the relevant academic disciplines—morphology, phylogeny, fossils, molecules—even as your irascible professors cling to their pet theories ad infinitum.

Fine-tuning the molecular phylogenetic tools is now revealing unexpected applications in areas that reach beyond taxonomy and species classification schemes, including the biotechnology revolution known as "molecular medicine." Phylogenetic reconstruction, the same tool we applied to the panda's roots, has proven invaluable in tracking reservoir species that harbor emerging viruses like HIV (from monkeys), Ebola (from rodents or bats), or the Serengeti lion's canine distemper virus (from hyenas). Phylogenetic approaches are being harnessed to characterize human gene families to trace the origins of hereditary diseases like Tay-Sachs, cystic fibrosis, and thalassemia, and also to evaluate aberrant gene expression in tissues and cancers.

Forensic applications using phylogenetic methods are even cropping up in the legal arena. In one bizarre case in 1994, a dentist in Melbourne, Florida, was accused of infecting his patients with the AIDS virus he carried. A sophisticated phylogenetic assessment of his and his patient's HIV strains proved he most assuredly had infected several of his patients, probably intentionally, before he died. Molecular phylogeny has been a key tool in tracking the source of the anthrax strains mailed by terrorists in the wake of the September 11, 2001 attacks on America. My NIH clinical colleagues are now mastering phylogenetic algorithms, developed to address taxonomic questions, but now so critical in biomedical research.

The most resounding message of giant panda biology, however, is that species conservation is an amalgam of scientific and nonscientific disciplines. To bicker over which is more or less important is a fruitless and time-wasting exercise. Many fields have staunch defenders, even disciplinary chauvinists, who would close their eyes and ears to other perspectives. This was never possible with giant pandas because every minuscule advance is broadcast by a vociferous media, always hungry for a morsel of giant panda news.

In 1996, a Chinese consortium announced plans to clone giant pandas as a conservation strategy. Was this a good idea? Pan Wenshi did not think so. He commented, "Pandas are not a test animal. . . . Other test tube projects have shown that for one successful test tube life, there are numerous failures. Do we have so many pandas to fail?" Added Lu Zhi, "There is nothing wrong with test tube technology itself. What is wrong is the choice of the panda as a candidate for the test." The controversies continue and we all want to help. Technology or biotechnology fixes are not the only way, nor always the best way.

Conservation biology is a young science. Early in the last century, bears, wolves, and wildcats were hunted to protect our children from what our forefathers perceived as deadly killers. Were Teddy Roosevelt, John Wesley Powell, and John Muir the first to worry about wildlife conservation at the turn of the last century? Had they read these words of Père Armand David composed in 1875?

From one year's end to another, one hears the hatchet and the axe cutting the most beautiful trees. The destruction of these primitive forests, of which there are only fragments in all of China, progresses with unfortunate speed. They will never be replaced. With the great trees will disappear a multitude of shrubs and other plants which cannot survive except in their shade; also all the animals, small and large, which need the forest in order to live and perpetuate their species. . . . It is unbelievable that the Creator could have placed so many diverse organisms on the earth, each one so admirable in its sphere, so perfect in its role, only to permit man, his masterpiece, to destroy them forever.

Ten

The Way We Were

Nothing in biology makes sense except in light of evolution.
—THEODOSIUS DOBZHANSKY

THE OLD TESTAMENT, AS INTERPRETED BY THE ESTEEMED seventeenth-century archbishop James Ussher, tells us that the earth is 6,005 years old and that Noah boarded his ark 4,348 years ago. As I grew up in the Christian religion of my parents and theirs, I was taught to accept on faith these and other biblical truths. Although my questions would be answered sincerely, there was peril in being too querulous, in not being convinced of the accuracy of the tenets of theological history. The danger was to be branded a heretic, an apostate, a pariah.

Scientists who study geology and the planetary history estimate the earth as a bit older, on the order of 4.5 billion years. Life started, they discern, not on a Monday of the Creator's busy week, but more slowly over millennia of chemical reactions in an oxygenated primordial soup that slowly oozed into a murky swamp of the earliest self-reproducing microorganisms. These primitive life forms were built upon a chemical code that specified a living blueprint, a DNA double helix of single nucleotide letters or base pairs with four basic varieties: adenosine-A, cytosine-C, thiamine-T, and guanosine-G. Two linear strings of millions of nucleotide letters on each side of a twisted ladderlike structure were attached by weak chemical bonds that ensured complementarity. The T nucleotide would always bind to A, and the G would invariably be matched by a C. (I used to remember this pairing as the initials of Thomas Aquinas and Geoffrey Chaucer.) So a GTAGTA stretch on one strand of the helix was matched with a CATCAT counterpart on the other. This A-T/G-C letter complementarity is the basis for DNA replication and of the primitive organisms' reproduction. That very simple chemical structure forms the platform upon which life on earth stands today.

The first signs of life appear in geological fossil remains that were deposited around 600 million years ago. Six hundred million years ago—how long is that, really? The trouble with these vast age estimates is that most people cannot imagine the difference between, say, 100,000 years, 1 million years, 100 million years, or 1 billion years. They all seem like a very, very, very long time ago! Yet trying to grasp what they mean can brightly illuminate the timing of biological species' comings and goings, including the how and why of our own human origins.

Fifty years ago a science journalist, James C. Rettie, proposed a graphic image of geological time in a popular magazine that was similar to today's *Reader's Digest* called *Coronet*. Rettie imagined a sophisticated space traveler from another galaxy who aimed a camera with a wide-angle super-telephoto lens at the earth before life began, some 750 million years ago. The alien photographer set the camera on time-lapse, taking a single image each year until modern times. Projected at the normal speed of twenty-four frames per second, the resulting movie, playing twenty-four hours a day, seven days a week, would take a year to view. Each day would span 2.1 million years and each month 62 million years. Here is what the movie showed.

From January to March, things are rather boring on the young planet, with little sign of life. The first unicellular microbes emerge in early April, and by the end of the month, multicellular aggregates have appeared. Late April and May are dominated by trilobite marine invertebrates, to be followed toward late May by the first vertebrate species. By July a panoply of colorful land plants are prospering and begin to blanket the globe. Late August brings amphibians, the first animals to make a cameo appearance on dry land.

In mid-September early reptiles preview the dawn of the dinosaur era, which will continue through late November, dominating the film and the world for seventy days. Birds and small mammals first appear in early November, but they are overshadowed by the diversity, size, and success of the dinosaurs. The giant reptiles overtake the earth, feasting on flora and fauna, soaring across the land and skies, exploiting available ecological niches and curtailing any advances by the tiny mammal precursor species under their feet. On December 1, the dinosaurs, all of them, abruptly disappear, victim to a global meteoric calamity that blackens the planet. At about the same time, the Rocky Mountains erupt on the western side of North America.

By mid-December some recognizable ancestors of modern families of mammals (e.g., monkeys, cats, bears, rodents, horses, etc.) begin to appear. Saber-toothed tigers come and go a few times in late December, but there are uncomfortably no signs of humanity. At midday on New Year's Eve, the earliest people finally make their debut with upright locomotion, enlarged braincases, and vocal communication. At 9:30 P.M., *Homo sapiens,* modern humans, migrate out of Africa to populate Eurasia and the Americas. At 11:52 P.M., lions, cheetahs, giant ground sloths, and mastodons become extinct in the Americas as the glaciers retreat for the last time. At 11:55 P.M., recorded human history and civilization as we know it begin. Twenty seconds before the end, Columbus sails to the Americas. Four seconds before midnight, the first automobile is invented.

In the movie (and in geological history) there are three periods of biological domination: the trilobites (April to May); the dinosaurs (September to November); and the mammals (November to December). The mammals flourish for seventy days of the movie, the first fifty days as rat-sized or smaller insect-eating creatures under the feet of dinosaurs. The last month sees their blossoming and domination of the earth when numerous mammal species hasten to backfill the ecological vacuum left by the dinosaurs' untimely exit.

Today, between 4,600 and 4,800 species of mammals, depending on how you count, live on earth. They occupy every continent and help define the ecological conditions they encounter. The morphological and adaptive differentiation among mammal species is enormous, ranging between the largest animal ever—the blue whale—to the echo-location-driven bats, to blind subterranean naked mole rats, and to cognitive humans. The richness of mammalian species' diversity and their remarkable specializations provide the evolutionary backdrop upon which the human species was created by formative evolutionary processes.

The script of our movie, indeed the source code for molding species by the miracle of embryonic development, is packed neatly in each cell of every tissue in living species. Tens of billions of copies of every gene, a written crypt of every vestigial mistake and each brilliant adaptive modification, all reduced to a linear tract of DNA letters, are carried on twenty or more (depending on the species) distinct chromosomes packaged neatly into sperm and eggs. And it works really well. Thousands of species happily replicate all the hard-learned lessons of history, thanks to their ability to compress and convey instructions in their tiny genome, the most intricate engineering blueprint ever to exist.

Genome science has finally developed the technology, resolve, and computational power to go after the decryption of living genomes. The striking parallels among the few mammals that we now understand in a genomics context lead to an indisputable conclusion: that genome scientists are studying variations on a single genome, that of our 100-million-year-old common ancestor. The diminutive mammal precursor species that scrambled from the dinosaurs of its day transferred a basic genome formula to all its descendants who would one day dominate the earth. Humankind grew from this same genomic lineage and carries with it a DNA platform and scaffold inherited from our primitive ancestors. Because of this evolutionary patrimony, the fits and starts, perils and successes of humankind recapitulate the natural and selected experiences of the mammal species of today, yesterday, and tomorrow. That is why animal geneticists absorb human genomics, and human geneticists ponder other mammals so carefully. To appreciate how intricately the human and animal genomes are joined by evolutionary forces, I need to describe first the disposition of our own human genome.

The Human Genome Project came into vogue about fifteen years ago. The effort involves an international collaboration of geneticists, government initiatives, commercial enterprises, biotechnology companies, and even ethicists, all attempting to unravel, read, and interpret for the first time the full DNA sequence of man and woman.

So what actually is a genome? A man's genome is the sum total of all his genes, his DNA, and his genetic information, neatly compiled in two distinct copies, one from each of his parents, in his every cell. Imagine the genome is an encyclopedia, the chromosomes are the volumes, and the genes are the paragraphs—some short and some long and complex. The three-letter codons, each specifying a specific amino acid used for the cell's protein synthesis machinery, are the words within each sentence. The letters in the words are the nucleotides or base pairs: A, T, C, and G. A complete sequence in each person's genome spans some three billion nucleotides, all in a cryptic language read by the translational (protein synthesizing) machinery of living cells. In all, one human genome encodes enough information to fill a hundred volumes of the *Encyclopaedia Britannica*. Genetic scientists are just beginning to decipher the rules for gene encryption.

In February 2001, the first full-length draft sequence of our human genome was deposited on a National Library of Medicine Web site (www.ncbi.gov) managed by the U.S. National Center for Biological Information in Bethesda, Maryland. The sequence represented the culmination of a Herculean effort to compile a complete "generic" human sequence. Rather than reading the sequences from one person (whom would they pick?), sequences from six anonymous volunteers were chopped by enzymes into bite-sized chunks, called molecular "clones," each about 150,000 nucleotide letters in length. These DNA clone bits were distributed to half a dozen DNA sequencing centers in different parts of the world who would each sequence thousands of DNA pieces. Powerful computer routines were applied to match overlapping ends of the sequences to string contiguous chunks together across each human chromosome. To ensure comprehensive coverage and also to minimize sequencing errors by the automated and very expensive DNA sequencing machines, each of the 30,000 to 40,000 molecular clones were sequenced ten times over. ·

The composite sequence was a monumental achievement for biology, for medicine, and for humankind. Looking back, it was only 150 years ago that Gregor Mendel, a German monk, launched the field of genetics by describing factors (called genes) that specified texture and color differences between strains of green peas. In the twentieth century, early geneticists studying corn and *Drosophila* (a tiny fruit fly) bred genetic variants and composed the first genetic maps. The genes that controlled eye color and bristle length in flies, kernel size and tassel texture in corn, were attributed to mutational variants arranged along tiny chromosomes like beads on a Catholic rosary. By the mid-1940s, DNA was shown to be the cellular chemical with which genes were built and transmitted to offspring. The self-replicating double helix structure for DNA was proposed by James Watson and Francis Crick in 1952.

The earliest human gene maps, locating hereditary diseases and variant proteins to chromosomal addresses, were built in the 1970s when the fields of gene cloning and molecular biology were in their infancy. The first gene to be sequenced into its exact nucleotide letters was a tiny alanine "transfer RNA" molecule, seventy-seven base pairs long, whose sequence in 1968 won Cornell's Robert Holley a Nobel Prize. By the mid-1980s, DNA sequence technology had improved considerably, and a decade later, DNA sequence assessment was automated. The initiative to determine the entire human sequence gained momentum in the early 1990s as DNA sequence reactions were delegated to robots and sophisticated biotechnology that ran around the clock, producing over a million base pairs of sequence in a single day. In 1998, the biotechnology giant PE-Applied Biosystems, which had developed the fastest automated DNA sequence technology, teamed up with Craig Venter, sequencing entrepreneur and genomic maverick. Their new futuristic genomic company, Celera Genomics, announced plans to tackle and finish the human genome draft sequence by 2001. That boast lit a candle under the seats of cautious but deliberate leaders of the international publicly funded Human Genome Project consortium. Within weeks, the public project would readjust their target dates to match Celera's. In February 2001, Celera's Venter and NIH genome project director Francis Collins jointly pronounced their private and public "draft" versions of the human genome complete and described how it looked in simultaneous dedicated issues of *Nature* and *Science*.

Achievement of the draft sequence was heralded as a milestone for genetics, for technology, for medicine, and for humankind. It was called the Holy Grail, the Book of Life, the Big Enchilada. Even disagreeable scientists who a decade earlier denigrated the mindless sequencing of the seas of "junk DNA" were coming around to the benefits of a complete sequence. The wonder of reading not just some but all of our genes, the entire unabridged set, had the ring of a brave new era for biology. Hereditary diseases, athletic and artistic talent, hair color, cosmetic appearance, even behavior can be attributed at least partially to genetic factors. Some envisioned a future where everything in biology that depends on our genes would be open for discovery. Here is what the Human Genome Project found in its initial look at the draft sequence.

First, only 1–2% of the three billion nucleotide letter sequence is part of a gene. Between 30,000 and 40,000 genes have been discovered in the human genome, neatly spaced in tandem along twenty-four unique chromosomes. No one is sure exactly how many genes there are altogether because when we read a naked DNA sequence, it is not so easy to recognize when a gene starts, ends, or continues— but we are getting better at this all the time.

We can identify the genes that we know encode proteins (e.g., hemoglobin, insulin, collagen, enzymes, and many more), but we only know the physiological role for about 8,000 genes. The other 25,000 are mysterious gene paragraphs awaiting a functional description. Also, genes are broken up into short coding DNA pieces called "exons," which are separated by short (or sometimes long) stretches of noncoding DNA sequences called "introns." Some genes have over twenty introns between the exons, while other genes have no introns.

To extend our analogy of a genome as a living encyclopedia, if the genes are the paragraphs, the exons are sentences and the introns are the commas, semicolons, periods, and exclamation points.

Adjacent to genes are regulatory sequence elements, stretches of DNA that determine in which tissue a gene turns on, how much of its product is made, and when to turn it off. These regulatory signals can be upstream of a gene coding exon, downstream, or even nestled within an intron. Then there is a lot of gibberish DNA between genes that probably does not do much; it is just there to hold genes together.

About half of the human genome sequence consists of repetitive sequences that occur many times in different chromosomal regions. The repeat sequences have different names depending on their size or frequency; they include LINE, SINE, minisatellites, and microsatellites. Many of the repeats are not even human originals, but DNA immigrants that have invaded our genomes from other species or microbes. A large fraction of these are endogenous retroviral sequences, similar to the *AKVR* gene described in Chapter 1. Such relicts of viral invasion descend from ancient epidemics and are carried in our genome as vestigial sequence prints. Normally they serve no useful function; they sit there like a genomic tonsil or appendix of little consequence to their present-day carriers.

The repetitive DNA sequences are mostly a nuisance to genome scientists because they make it difficult to match up overlapping DNA chunks (the 150,000-base-pair molecular clones that get sequenced). But once in a while they can be useful. For example, the tiny repeat sequences called microsatellites, markers we used to estimate the dates of the population bottlenecks in wildcat species, are also extremely variable among people. Microsatellites are short, two-to-five-nucleotide-letter repeats that appear in DNA sequences every 10,000 nucleotides or so. Because of the extraordinarily high variability among people, the 100,000 randomly spaced microsatellites have become a favorite marker for mapping genes because they are easy to follow in family studies. Microsatellites have also proved to be powerful tools for the forensic community in matching blood or semen samples left at crime scenes (as we will see in Chapter 11).

A critical discovery that came out of the Human Genome Project involves the enormous amount of DNA variation that is dispersed across the human genome. It turns out that a common nucleotide letter variant occurs about every 1,200 nucleotide letters of sequence. This means that around 5.0 million single nucleotide polymorphisms (SNPs) exist among different people. Some of these occur within coding exons of genes, others in regulatory sequence elements, but most fall in the sea of noncoding DNA that composes 98% of our genome. SNP variants have been shown to be the principal cause of inherited genetic diseases. Bad SNP mutations in good genes are responsible for hereditary ailments like cystic fibrosis, sickle-cell anemia, muscular dystrophy, and Tay-Sachs disease. But the vast majority of SNPs have not been connected or associated with human character differences. Some of us feel that the biggest promise of the Human Genome Project is to connect the numerous individual SNPs to human differences in disease, talent, appearance, and behavior.

So now that we have a glimpse of the full human sequence, what good is it? What does it mean? Does it make life better or worse? Plenty of ink has been spilled answering these questions, and the answers are all different. Eric Lander, director of MIT's large human genome sequencing center, perceives our genome sequence as a "parts list," comparing it to a catalogue of 100,000 pieces accumulated from a disassembled Boeing 727. He muses that we would have trouble putting all the parts together given the collection and no manual. Even if somehow we did, it probably would not fly! City of Hope geneticist Susumo Ohno looked at the human DNA sequence and transposed it into a musical composition, which he played on a Yamaha synthesizer at his lectures.

Medical doctors see the genome sequence as the first step toward identifying the over 2,000 hereditary diseases that afflict us. Earlier, more accurate diagnosis of genetic maladies based on pinpoint mutations would allow exact genetic counseling and preventive care. Mutation assessment of human hereditary diseases would unleash innovative therapies, attacking or compensating for the lost function of defective genes.

Oncologists look forward to genetic applications in accurate diagnosis of the hundreds of human cancers and "smart bomb" treatments that home in on tumors in specific tissues, all the while leaving normal cells unfettered. Gerontologists dream of understanding aging better and ameliorating its debilitation. Psychiatrists hope the new genetics will make psychological diagnosis and therapy more objective.

A prospering biotechnology industry has spawned over patent licensing of human genes, their variants, and their therapeutic potential. Pharmaceutical research and development laboratories are looking to connect SNPs in drug metabolism genes to some people's adverse or beneficial reactions to new pharmaceutical compounds. The day will come when general practitioner physicians will assess each patient's key genes prior to prescribing medication. Newborn babies will be sent home with a DVD or compact disc annotating their genotype for thousands of their genes. For patients it will mean an end to the trial-and-error drug treatments so common in our over-medicated society.

My own perception of the human genome sequence is more akin to a rediscovered long-lost diary of an intrepid explorer written in a mysterious long-forgotten language. Nestled in every chapter are discoverable clues to humanity's past, relic genetic pawprints of the skirmishes our ancestors engaged in for survival. Large and small success stories are all there, ratcheted together as accumulated trophies of ancient engagements. Each chapter, verse, and volume holds the secret records of our ancestors' cumulative encounters with extinction and adaptation, patiently awaiting our primitive attempts to decipher messages that unveil the fiber of our existence. Along with the recipes for countless success stories are notations of the false starts, dead ends, and mutational missteps, new strategies for species and individual propagation that proved less suitable than what came before. To decipher this exquisitely elaborate code would be to rediscover the origins of humans, of animals, of life itself.

Unlike the decryption of more familiar ancient codices—the Dead Sea Scrolls or the Rosetta stone—there exists not one, but virtually unlimited copies of the human genome. To be exact, there are two in each of the six billion people on earth, easily accessible from a cheek swab or tiny blood sample. But when the experts read the sequence, they do not really understand what a lot of it means. Genes are not always obvious, and when they are, it is not clear what they do. Regulatory sequences, the dimmer switches for genes, come in many varieties, most of which are not well understood. Genes cluster very closely in some chromosomal regions, while in other areas they are scattered across vast deserts of nonsensical DNA devoid of genes or anything we can recognize. A critical strategy to interpret genome sequence will surely come from an unexpected source, the comparison of human genome patterns and organization with the genomes of other species—the contemporaneous blossoms of evolutionary radiations.

The comparative approach is not new to biology; it has been used for centuries. Our study of human anatomy and function has benefited immensely from comparative anatomy, that is, by studying organ metabolism and differences among cats, pigs, and monkeys. Comparative physiology of laboratory animals greatly informed human renal, heart, muscle, and reproductive physiology. The same is true for biochemistry, neurology, and immunology. Now is the time for comparative genomics.

Comparative genomics represents a paradigm shift, a reversal in how biological inference is advanced. For the other comparative disciplines, differences were interpreted as specialization or adaptation for the ecological niche, habitat, or lifestyle a species chose. Land animals use lungs to capture oxygen, fish require gills, and certain lizards breathe through pores in their skin. For genomics, we are comparing not the functional structure, but the gene script for all such adaptive functions and anatomical modifications at once. The genes we compare and their accumulated changes determine every comparative variation that came before. Then by detecting the patterns of natural selection, we deduce the origins and mechanisms by which our genes evolved, and how they work.

Because of the promise of comparative genomics, there has been a revived enthusiasm for gene mapping and genome sequencing in other species. We start by building detailed gene maps or full-length genome sequences of vastly distinct species and then examine how their homologous genes and genomes are organized. First to be considered for full genome sequence were the traditional "model" organisms for genetic study, *Escherichia coli* (a bacterium), *Saccharomyces* (a yeast), *Drosophila* (a fly), and *Caenorhabitits elegans* (a roundworm). Besides humans, two mammal species, mice and rats—long-standing medical models—have already been selected for full genome sequencing. In December 2002, a very high quality draft sequence of the mouse genome was released by an international sequencing consortium. In the wings are plans to sequence a dozen other mammal species, each of which has taken the ancestral mammalian genome in a different direction.

Back in the 1970s when human gene maps were being assembled for the first time, smaller cottage industries to map other species sprung up as well. The agricultural community yearned for gene maps of barnyard animals—cattle, sheep, and pigs—so that economically important traits that influenced size, leanness, and disease resistance could be managed in large herds. Chicken researchers began to develop a map of that species, looking for a genetic edge to better fryers in the huge food industry.

Genome mapping projects for chimpanzees have grown to calls for all the genes of humans' closest relative to be sequenced. When we look at samplings of chimpanzee DNA, its gene sequences are on average 98.5% identical to human DNA. To help grasp how very closely related we are to chimpanzees, British science correspondent Matt Ridley, in his excellent book *Genome,* posed a colorful analogy. Suppose George W. Bush stood in the White House in Washington and held hands with his mother Barbara. Then Barbara held her mother's hand and she held her mother's and she hers and so on. As the line of mothers stretched out New York Avenue and up I-95, by the time the human linkage reached New York City, the American president would be holding hands with a chimpanzee.

Today's proponents of a full genome project for chimpanzees believe that by comparing a chimp sequence to a human sequence, we will begin to understand the basis for human development of upright walking, cognitive thinking, and complex language. Japanese and American scientists announced in 2002 plans to begin sequencing the chimpanzee genome in the next few years.

By the mid-1970s, it was evident that the comparative genomics approach had limitless potential for revealing evolution's boundless creativity. I also reckoned that it had a real future in comparative medicine, but I needed to select a mammal to map. With specific designs on biomedical applications, I picked the common house cat.

I chose domestic cats first because they were afflicted with feline leukemia virus, FeLV, a retrovirus that causes leukemia to act like an infectious disease. I suspected that the interaction of FeLV with the cat's immune response genes would be a fertile area to study. My hunch was realized as studies with FeLV revealed how the virus would integrate its tiny genomes into random positions along the chromosomes of a cat's white cells. When they parked adjacent to a group of genes we now call oncogenes (because they can cause cancer), the powerful FeLV regulatory sequences switched on the oncogene's expression, causing the cell to divide itself uncontrollably, the hallmark of leukemia. Over a hundred different oncogenes have now been discovered via FeLV in cats, and their homologous counterparts were identified in mice, chickens, and humans. Today human oncogenes are the focus of cancer diagnostics and suitable targets for exquisite designer-drug therapies. Identifying the precise oncogene that misfires in different cancers is turning around the hopelessness of cancer, because they have directed researchers to the specific details of how cancer really works.

As our research group persisted in building a feline gene map, we uncovered other benefits to mapping the cat's genes. A powerful advantage lies in their medical scrutiny. There are twenty-seven veterinary schools in the United States, each producing perceptive clinical practitioners who treat our pets, over 100 million cats and dogs. Veterinarians assess and interpret animal disease and then publish their findings in a vast and growing veterinary literature, drawing heavily on comparative similarities to homologous human disease.

Today, some two hundred human hereditary diseases have counterparts in the cat. Diagnosis and treatment of these feline diseases have advanced considerably thanks to human research advances. Likewise, research on the feline genetic disease has become quite informative in revealing the physiological basis of the homologous human diseases. Also, many if not most of cat hereditary disease models simply do not exist in more widely studied mice and rats.

Viruses and other microbial agents of the domestic cat have interesting biomedical parallels as well. Feline immunodeficiency virus (FIV), feline distemper virus, feline calicivirus, and feline infectious peritonitis virus (which afflicted the cheetahs) are all devastating microbes that have revealed grim details of infectious disease onset and progression. In the mid-1970s, Cornell virologist Colin Parish showed that the feline panleukopenia (distemper) virus cultivated in a cat vaccine factory abruptly jumped from cats to become a hypervirulent strain in the world's dogs. That virus transfer fostered a global epidemic of neurological distemper disease that went on to kill millions of puppies before a canine vaccine was developed. The ghastly parallels of abrupt virus outbreaks would repeat itself in humankind with flu, AIDS, Ebola, West Nile, and mad cow disease, each originating in animal species.

The cat was an attractive species for which to construct a gene map for two additional reasons. First, the domestic cat is a member of a family of wild cats, Felidae, whose many wonderful species are easily accessible in zoos. That may sound trivial, but if you try to assemble blood samples from the twenty species most closely related to a mouse or rat, you will spend a lot more effort catching them than I do with the big cats. Second, because pet cats are domesticated, there exist today some forty recognized breeds that have been artificially bred to display different colors, textures, shapes, and dispositions. We set out to pinpoint the genetic variants that specify as many hereditary variations, medical and otherwise, as we could in the cats.

From the start, we mapped the cat's genes, enzymes, oncogenes, and proteins with biochemical functions. Using tools similar to those developed for human gene mapping, we gradually assembled a feline gene map with hundreds of genes. They were placed on specific positions on chromosomes in different ways. One method, fluorescent in situ hybridization, involved tagging molecular clones isolated by molecular biologists from human or mice genes with a colored dye, and then adding them to a cellular array of cat chromosomes spread apart on glass microscope slides. The brightly colored gene pieces would seek out the chromosome position where its cat counterpart was located and form a complementary DNA helix in a snap-match reminiscent of a Velcro zipper. In this way, we could identify the precise chromosome position for the gene. Another gene mapping approach involved tracking several gene variants together in pedigrees of matings between cats. Genes close together on a chromosome are "linked" and will be transmitted together to kittens most of the time in these pedigrees.

Eventually we had mapped enough feline genes to compare the cat's arrangement to the gene map of humans. At last we were ready to take a look at the differences and similarities between the genomes of the human and the cat, species who we knew last shared a common ancestor some ninety million years ago.

To appreciate that comparison, I should first explain the relationships between human and primate genomes. Before the more powerful tools of molecular biology and human genomics were invented, cytogeneticists used the Giemsa chromosome dye to define a characteristic banding pattern of human chromosomes and those of other mammalian species. What they discovered is that the human karyotype (the sum view of a cell's chromosomes) had twenty-three chromosome pairs, each with a distinctive banding pattern, a sort of linear "bar code" pattern for every chromosome. The chromosome banding patterns of humans' closest relatives—chimpanzees, gorillas, and orangutans—were amazingly similar; only a few ape chromosomes showed evidence of rearrangements—that is, breaks and rejoining elsewhere—compared to human chromosomes. The chromosomes of more distant primate species—baboons, vervets, mangabeys, and other Old World monkey species whose common ancestor with humans/apes reaches back to around twenty million years ago—were more different from humans than the chimp's or gorilla's, but only slightly. Chromosomes of the New World monkeys of South America (owl monkeys, spider monkeys, and marmosets) were also quite similar in their appearance to humans, even though some forty million years have passed since the common ancestors of New World monkeys and apes walked the earth.

Early gene maps of several primate species showed that genes that were linked together in humans were almost always strung together in other primate species in the same order. Very few exceptions occurred. A new procedure called "chromosome painting," which illustrates chromosome sequence homology by direct observation, brilliantly confirmed the parallel organization of the genomes of primates. In that method, individual human chromosomes are purified and separated by differential refraction in a laser beam by an expensive machine called a "fluorescent activated cell sorter" (FACS machine, pronounced the same but invented before the familiar fax machine). Then DNA from a single individual chromosome is sheared into small pieces and chemically labeled with a fluorescent dye. The labeled human chromosome's DNA, called a "probe," is added to the chromosomes of another species (e.g., a baboon) spread apart on a microscope slide. The labeled human DNA finds its homologous DNA counterpart among the baboon chromosomes, hybridizes, again like a Velcro zipper, and colors the chromosome stretch with the bright dye. This allowed us to see large segments of the baboon chromosomes as precise evolutionary matches for the human chromosome we had labeled. By repeating the experiment using each of the human's twenty-four chromosomes we were able to develop a chromosome-by-chromosome snapshot of the baboon's linear gene arrangement compared to that of the human genome.

So far twenty-five different primate species have been painted using the human chromosome probes. As we predicted from the gene map parallels, these paintings reveal an extremely high degree of genomic conservation. Most primate species show between eighteen and twenty chromosomes identical to the human version. The few changes that do occur involve single breaks of three or four or five chromosomes. By comparing Giemsa banding and paint results for these species, the genome organization of the long-forgotten common ancestor for today's 280 primate species was deduced, even though that creature ceased to exist over sixty-five million years ago. The ancestor to all the primate species enjoyed a master-genome organization that has been retained almost completely intact in all its descendants. Fewer than eight exchanges between different chromosomes have occurred over the evolutionary history of the order of mammals we call primates.

The cat is one of thirty-seven species of the family Felidae, itself one of seven families in the mammalian order Carnivora. Carnivora (350 species) includes cats, mongooses, civets, and hyenas as one subordinal lineage, and bears, dogs, skunks, weasels, and seals as the other. Like primates, the domestic cat genome is also highly conserved among other wild cats. Sixteen of the cat's nineteen chromosome pairs occur unchanged (i.e., by chromosome exchange) across the other Felidae species. Further, each of the sixteen invariant Felidae chromosomes are intact in other Carnivora species. So Carnivora genome evolution is also very slow and nearly invariant since divergence from their earliest common ancestor. Like the primates, there once existed a master ancestor from which modern Carnivora genomes descend. By now we have developed a pretty good idea what that first ancestral Carnivora genome looked like. It looked a lot like the genome of our pet tabby.

So what happened when we compared the cat genome to the human genome, each one showing powerful resemblance to the primitive ancestral genome arrangement of its own mammalian order? You guessed it. The human and feline genomes display extraordinary similarity. Long strings of genes, sometimes extending over entire human chromosomes, are recapitulated precisely with the same (homologous) gene arrangements in the cat. Cats have nineteen chromosome pairs; humans have twenty-three. But chromosome after chromosome, the same genes are hooked together in cat as they are in human. Only a handful of exchanges between chromosomes, around ten, account for the differences between chromosome content in the genomes of cat and human. If we ignore the within-chromosome flips in the cat and human genome organization, it takes only ten scissor-snips to rearrange all the feline chromosomes into the human genome organization.

Genome conservation spreads between other mammalian orders as well, dating back to splits that occurred before the dinosaurs disappeared, eighty to ninety million years ago. Comparative gene mapping and chromosome painting have recently been used to show that the conserved ancestral genome organization is also apparent when we inspect diverse species including rabbits, lemurs, cows, pigs, whales, horses, bats, common shrews, and tree shrews. The master rationale and patterning for the genome organization found in all mammals is close to being solved. The genome of the mammalian ancestor looks remarkably similar to that which specifies human and cat, two species that retained intact the major genomic features of their very old common ancestor.

The remarkable conservation of the mammalian genome has some important exceptions. Mice and rats have their genome appreciably reshuffled as compared to the human-cat-ancestral disposition. The murine genome is broken up into approximately two hundred short chromosome segments that correspond to homologous human segments. The same is true for the rats. This is nearly five times as many chunks as we see in comparing human to cat. In the rodent evolutionary history, a global genome reorganization must have occurred, one that is reflected today by the patchwork reassembly that appears when we compare rodent and human genomes.

At least two similar genome shuffles also occurred during the primate species radiation, one on the branch leading to lesser apes or gibbons, and a second during the ancestry of certain New World monkeys, particularly owl and spider monkeys. Among the carnivores, the dog family Canidae and the bear family Ursidae are likewise punctuated with global genome reshuffling relative to the primitive ancestral carnivore genome disposition. The rapid genome rearrangements in the bear family provided important genomic signature patterns that helped us work out the ancestry of bears and the giant panda.

So it seems genome evolution in mammals has proceeded at two very distinct rates: the very slow or "default" tempo that allowed us to reconstruct the ancestral mammal's genome organization, and a more rapid global reshuffling apparent in the exceptional genomes of rodents, gibbons, dogs, and bears. The slow and steady rate of genome evolution is found across all the mammal orders, but it has been punctuated on certain rare occasions by a genome-wide reshuffling event that persists today in descendant species of those episodes.

The cause or driving force for the periodic reorganization is a mystery to genome scientists. Some observers wonder if repeat sequences in our genomes, particularly mobile elements like the endogenous retroviral components, might predispose chromosome breakage. This is assuredly the case in several insect species where the repetitive sequences, relics of viral infections, predispose their host's genomes to shattering and rearrangement. The proof of this contention in mammals is less certain, but the footprints of these elements will soon help resolve this question.

Fulfillment of the hopes and fears of the post-genomic era may seem a slow process for those of us involved, but it will appear breathtaking in its rapidity to our alien movie photographer. The annotation of the human genome will identify and ascribe physiological functions to many of the more than 30,000 genes and place the rest in related gene families. Over 1,000 human hereditary diseases have already been connected to single nucleotide variants. Shortly, that number should double or triple.

Developing gene maps of pet and livestock species will gain steam and emerge as tools to map economic or disease-related gene differences, all the time offering comparative genetic details for the human situation. Information flow from mouse/rat models to human will be expanded to include dogs, cats, and farm animals. Veterinary diagnosis and treatment will blossom. Veterinary medical advances will inform and expand human treatment regimens based on research on the 500 cat and dog hereditary diseases with human homologue counterparts. Genome maps of animal species will guarantee their utility and application as medical research models. Veterinary schools will become more active centers for medical genetics because the animals have untapped biological diversity. Full genome sequencing initiatives will reach a high point as a new mammal's genome is sequenced every year and then perhaps even more frequently.

Gene handling, gene regulation, and gene control in animal models will answer many functional questions about the sea of genes for which we have identified no function, no name, no clue. Comparative genomics will develop as a critically powerful tool for human gene annotation. The phylogenetic analysis of gene families within and between species' genomes will permit the implication of historic events that mold and favor disposition of modern genes.

Evolutionary biologists trying to explain the forces of speciation, survival, and extinction will have a new tool—gene maps—and a new charge: to pinpoint the adaptive changes in genes that mediate the fate of a species. Geneticists have traditionally depended on selectively neutral DNA variants—allozymes, microsatellites, and noncoding DNA sequences—to track the divergence history of species. The precise gene alterations that contributed to a species' physiological adaptations are still largely unknown. Not for long. Now the genes that capacitate survival and adaptation of living species will have a chance for resolution.

Not since the earliest paleontologist realized the importance that fossil deposits would have for revealing our past has so much potential been anticipated. Comparing genome organizations is a new treat in our search for our deepest biological roots and for the recipes for survival. As we draw this picture into sharper focus, as gene maps get cleaner, we are decoding the most complex and exquisite cryptographic pawprint ever developed: the linear gene assemblage that specified the mammalian evolutionary radiation. Some rather cogent illustrations of the translational benefits to the human species offered by our animal cousins follow in the coming chapters.

Eleven

Snowball's Chance: Genomic Pawprints

THE SETTING COULD HARDLY HAVE BEEN MORE BEAUTIFUL and serene. A Technicolor landscape of rolling meadows and rich forest, this tiny island province in far-eastern Canada was a vacationer's dream. Prince Edward Island (P.E.I.) is best known to non-Canadians as the setting for the gothic romance tale *Anne of Green Gables*. P.E.I. relies heavily on tourism, and once the vacationers depart in autumn, the native populace returns to their families, their faith, and their struggle to make ends meet.

By the summer of 1994, Shirley Ann Duguay was having a rough time in rural P.E.I. The pretty, slight, thirty-two-year-old single mother of five children was struggling on welfare. With emotional support from her extended family, Shirley clung to small threads of hope for a better future. She had made the decision to terminate her twelve-year on-and-off common-law relationship with Doug Beamish, father of three of her children, and when a free moment became available she began seeing a handsome young fisherman, Alfred Casey, from the next town.

Shirley and Doug had endured a stormy relationship across the years. They separated every four or five months, generally when his temper led to a physical beating, only to reconcile shortly thereafter—not an unusual pattern for desperate domestic abuse cases. But that dark chapter was over forever, Shirley thought. She had told Doug once again of her firm resolve to sever their union one Saturday night in October, leading to a loud shouting match overheard by

Shirley's young neighbor, Linda Ranier, who was staying the night with Shirley while her family was out of town. Beamish departed, climbed in his pickup, and took off. On Sunday, October 3, 1994, Shirley Ann Duguay disappeared from her home as Linda slept, never to return.

The police were notified, but they did little for a few days. Later, they would admit that they suspected foul play from the start, but they were hesitant to comment further. Four days later Duguay's 1982 Buick was found abandoned in the woods outside Tyne Valley, ten miles from her home. The windshield and seats were spattered with bloodstains, lots of them. The police took blood swabs from the car and sent them to the Royal Canadian Mounted Police (RCMP) forensic laboratory in Halifax to be identified. The blood came from Shirley Ann Duguay as imputed from the DNA-based genotypes of her children.

Volunteers from Richmond, Summerside, and throughout the area mobilized to search the region for clues to Shirley's disappearance. They combed the area around her home and where her vehicle was discovered. Within a week sixty cadets from the provincial police academy joined the search, determined to find evidence of Shirley's fate. Grievous violent crimes were quite uncommon in P.E.I. and everyone was haunted by the unspeakable prospect of what might have happened.

Three weeks after Shirley's disappearance, the island residents were in an emotional turmoil, rife with fear, suspicions, and conspiracy theories. The Canadian army dispatched a squadron of 150 infantryman to search the wooded areas around the house, the car, and locales suggested by telephone tips. From dawn to dusk the soldiers marched shoulder to shoulder, combing for evidence and for Shirley Ann Duguay. Daily television updates chronicled optimism gradually fading to pessimism and fatalistic acceptance of the terrible prospects. Six weeks after her disappearance, the trail had gone cold. No suspects, no body, no evidence—another unsolved missing person.

A squab farm near Lake Casitas, California, with a hidden infestation of mice carrying a golden nugget in their genes. *(Murray Gardner)*

The African cheetah is the world's fastest land animal. *(Karl Amman)*

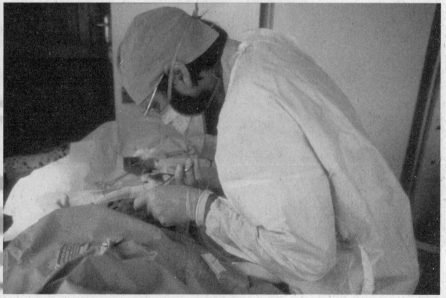

Mitch Bush performs a surgical skin graft on an anesthetized captive cheetah in South Africa, 1983. *(Stephen J. O'Brien)*

The genetic research team samples lions in the Serengeti to detect genetic impoverishment and infectious disease. The lions are asleep and will soon awaken. (*Stephen J. O'Brien*)

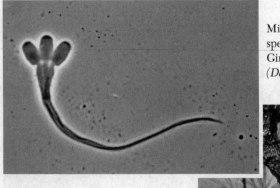

Microscopic view of abnormal sperm observed in semen from Gir Forest Asian lions. (*David Wildt*)

A Florida panther treed by Roy McBride's dogs in the Big Cypress Swamp in South Florida. (*Melody Roelke*)

A mobile lab in the alligator- and mosquito-infested Florida swamp used to gather precious fluids from a Florida panther. (*Melody Roelke*)

Scott Baker holding a crossbow aboard the rubber Zodiac raft and looking to collect a skin-punch biopsy from a passing humpback whale. (*Stephen J. O'Brien*)

A lion suffering from a fatal wasting disease that killed a third of the lions of the Serengeti population in 1994. *(Melody Roelke)*

Dr. Melody Roelke and her Tanzanian intern examine a lion in East Africa during the medical alert in 1994. *(Seth Parker)*

Junior, an adolescent orangutan son of a Bornean and Sumatran orangutan marriage, living at the National Zoo, Washington, D.C. (*Jesse Cohen*)

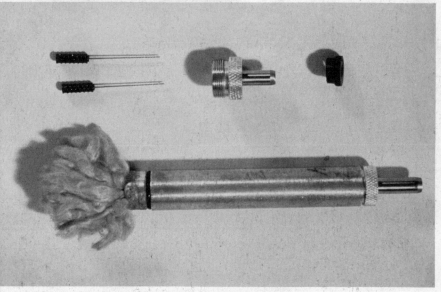

A skin biopsy dart designed by Dr. Billy Karesh for orangutan sampling. (*Billy Karesh*)

A young giant panda relaxing atop his habitat in the Qin Ling mountains in China. (*Lu Zhi*)

Research fellow Lu Zhi carefully plucks a hair for genetic sampling from a cub in her den in Shanshuping Forest Reserve in China. (*Stephen J. O'Brien*)

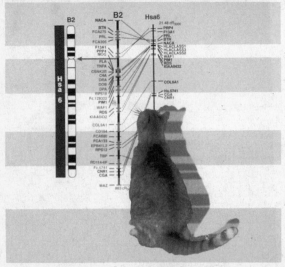

A young cat inspects her genome for the first time. (*Ellen Frazier*)

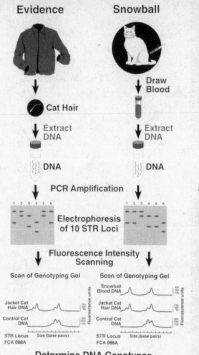

A graphic display used to explain forensic DNA typing of Snowball, the cat who tied his owner to a murder scene. (*Marilyn Raymond*)

The AIDS quilt displayed across the United States Capitol Mall in Washington, D.C. (*Larry Morris*, The Washington Post)

LENTIVIRUSES

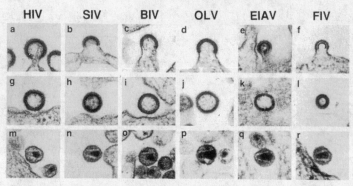

Electron microscope image of HIV and other genetic relatives in monkeys (SIV), cattle (BIV), sheep (OLV), horses (EIAV), and cats (FIV) (*Matthew Gonda*)

Jesse Gelsinger on holiday in Philadelphia before he became the first fatality in gene therapy clinical trials (The Washington Post Magazine, *December 20, 2001*)

David Wildt examines a semen sample in Tanzania's Ngorongoro Crater on the tailgate of a Land Cruiser dubbed NoAH, as the sperm donor looks on. *(Stephen J. O'Brien)*

The telephone rang in my office late in the evening that December, long after my secretaries had gone home. I answered it hoping it was not a scientific journal editor prodding me to supply a manuscript review I might have promised weeks or months before. The caller introduced himself as Constable Roger Savoie of the RCMP. He asked if I would be willing to take a few minutes for him to explain a missing person's case, a possible homicide, that he was investigating. I told him to go ahead.

Savoie outlined the background of Shirley Duguay's disappearance. He mentioned that his prime suspect was Douglas Leo Beamish, the ex-boyfriend who had a prison record and an anecdotal reputation for being rough with his girlfriends, including Shirley. Savoie explained the absence of a body, the presence of Shirley's blood in her abandoned car, and the military search. The soldiers had come up with a bit more than had been publicly disclosed. On their third day, they discovered a plastic Canadian tire bag in the woods. Inside was a man's leather jacket and running shoes. Both were spattered with blood that by DNA tests matched that of Shirley Duguay, but no other person's DNA was found. In the lining of the jacket they also found a whorl of long white hairs. But when the forensic lab in Halifax looked at these, Duff Evers, one of a dwindling breed of forensic specialists in hair identification, concluded that the hairs came from a cat and not from a human.

Savoie was disappointed, because he was hoping the hairs might be from the murderer. Then a flashbulb jolted his memory. When he had first questioned Beamish at his parent's home, Savoie noticed a rather large fat white tomcat, introduced to him as Snowball by Beamish's parents. Beamish had been living at his parents' since his recent release from prison.

"Dr. O'Brien," he continued, "I have been searching for weeks for a forensic laboratory that can tell if, or better prove, that the hairs from the leather jacket came from Snowball." Human forensic labs had repeatedly declined. Some even thought his request was laughable, a joke. Their reason: All the forensic advances in genetic identification were based on human genes, human genetic signatures, human DNA fingerprints, and never genetic "pawprints." Nobody did this for cats. It might be possible in theory, but the startup and background would take much time, be very costly, likely prohibitive. Further, feline DNA genetic matching had never been introduced in a murder trial. In fact, Prince Edward Island had yet to see human DNA evidence used in a homicide or any other criminal case.

Savoie explained that he had found my name on the Internet and then checked out my professional reputation with a few forensic geneticists. He pleaded that I was his last hope and that he would gather up as much financial means as the RCMP and crown prosecutors' office could afford.

Would I be willing to type DNA from the hairs in the jacket and compare them to Snowball, perhaps closing the circle of evidence by linking cat DNA from Shirley Duguay's blood to the home of the principal suspect, Doug Beamish? As he was speaking, I thought to myself, "Now this is really interesting!" So I answered back, perhaps too quickly, "We can sure give it a try!"

The next day, I summoned Marilyn Menotti-Raymond and Victor David, two crack researchers in our group, to my office. The two were steering our feline genome project. Marilyn had been an unusually bright graduate student who, though chronologically as old as I, had the youthful exuberance and research enthusiasm of her much younger colleagues. Her dad was a founding scientist of the Bristol Myers-Squibb pharmaceutical company. She had married quite young and raised two handsome sons before returning to graduate training in molecular biology and genetics in her early forties. She matched the burning scientific curiosity of fledgling young apprentices to the deliberate care and maturity of her more senior scientific colleagues.

Marilyn knew a lot about the feline microsatellite markers, the nucleotide letter stutter repeats so useful in human gene mapping. Indeed, she and Victor had isolated and characterized over four hundred feline microsatellite marker loci, the sites on a chromosome where a microsatellite is found in the cat's genome. These were the same genomic markers we had used to unravel the natural history of the cheetah and the Florida panther. Because microsatellites were so variable in humans, they had rapidly become the favorite genomic currency for forensic identification. Marilyn and Victor had been very busy locating microsatellite markers on the cat genetic map when Constable Savoie had called.

At first Marilyn was dubious and reluctant; there was too much background work to consider. Select the best microsatellite markers; make sure they work; get them from cats, then from cat hairs. And what if they did match? The human forensic community had spent fifteen years wrestling with DNA fingerprint efficacy and interpretation with help from the best minds in human population genetics. We were just one small laboratory, a cottage industry driven by youthful idealism and a fondness for cats. This was no match for the uncertainties, responsibilities, let alone the viciousness, of homicide trials.

I knew the case would be a challenge, but I was also certain we could do it. My background in population genetic theory and practice, the stuff of forensics debates, would help. The National Research Council (NRC), an advisory arm of the National Academy of Sciences, was about to come out with their second report on DNA-based forensic applications. The first one had appeared in 1992. These reports were how-to manuals for the rigors of forensic genotyping. We alone had the feline microsatellite tools; nearly all were unpublished. The intrepid Mountie had implored that we were his last hope. If anyone could solve this case, it was us. Victor agreed.

We also had a few friends we could consult who were more experienced in forensic genetics. My friend Bruce Weir had analyzed the DNA evidence in the O.J. Simpson trial and was a giant in population genetics theory. We knew we could call on Victor McKusick and James Crow, principal authors on the 1992 and 1996 NRC reports on DNA technology in forensic science. And we had Lisa Forman, lead DNA technology practitioner at Cellmark Diagnostics, the foremost company supplying DNA typing in high-profile forensic cases. In her youth Lisa had been a postdoctoral fellow in our lab studying the genetics of South American golden lion tamarins, and we were still good friends. She had testified at scores of murder trials on the power of DNA typing.

Later that morning I telephoned Lisa Forman, explained our predicament, and asked for her advice. Her initial reaction was incredulity, but as I went over the details and registered my optimism, she warmed to the idea. Lisa was adamant about the extremely meticulous care we should take with our research—take faultless notes, keep samples in a locked refrigerator, label and double-witness each move. The weakest point of any case was what she called the "chain of evidence"—keeping track of every single step in handling physical evidence (i.e., the jacket hairs) from crime scene collection to lab analysis to return to law enforcement officials. Break the chain or simply fail to prove the chain was unbroken, and lose the case. Be overly fastidious in every step and be certain the results are solid enough to prevail in the face of harsh, destructive, *ad hominem* assault in cross-examination. As Lisa detailed the forensic requirements I could feel my own resolve slipping, but I never let it show. I had unusual confidence in the meticulous and precise care that was standard for the Marilyn-Victor team.

That afternoon I spoke again with Constable Savoie. Yes, we would try to help, but we made no guarantees. I instructed him first to find a competent veterinarian and then obtain a subpoena from a judge to collect a blood sample from Snowball. Go to Beamish's house, serve the subpoena, have the vet draw the blood. Tell no one, not even the vet, what this was all about. Let them think you are going fishing or maybe just nuts.

I told Savoie to place the blood in a canister, wrap it with evidence tape, and put it in the refrigerator. Lisa's chain of evidence lecture persuaded me not to even consider Federal Express–like couriers, so I told the constable to put the white jacket hairs in a second canister and to book the next available flight from Prince Edward Island to Washington-Dulles Airport.

On January 4, 1995, Marilyn drove to the airport to collect Constable Savoie and two canisters of evidence hand-carried on his person. Marilyn, an incurable romantic, had imagined the Mountie would be a tall, dark, and handsome uniformed image of Nelson Eddy. Savoie was not so tall, a bit chubby, and disheveled in his jumper, jeans, and sneakers as he disembarked the plane. He clutched the evidentiary hair and blood as he rode to our NCI-Frederick laboratories with Marilyn. He told her how the customs officer in Boston had asked about the canisters. When he explained they were cat hairs, the official demanded they be opened. Savoie said that he would be happy to comply, but she should then be prepared for a Prince Edward Island subpoena in a murder trial coming up. She waved him through.

Marilyn and Victor documented receipt of the evidence from Roger and padlocked the materials in separate refrigerators—one for the hairs, one for Snowball's blood samples. Roger Savoie stayed for less than a day, but his determination, conviction, and attention to detail convinced me that the citizens of P.E.I. were being well served. I was actually a bit proud to be part of his effort and crossed my fingers that we might successfully extract DNA from the jacket hairs.

Victor and Marilyn had developed their own personal laboratory tutorial on the practice and pitfalls of DNA forensics. First, they designed preliminary experiments to search for the microsatellite markers that were most sensitive to tiny amounts of cat DNA. They then practiced employing the powerful polymerase chain reaction (PCR), which Scott Baker had used to copy whale DNA from Japanese sushi in Tokyo hotels, to amplify a few molecules of a cat's DNA so it could be typed for microsatellite alleles. Once the microsatellite methods were ideal, Victor and Marilyn fiddled with their sensitivity using trace DNA specimens from cat hairs, blots on filter paper, T-shirts, and mixes of different species. All the while, Savoie and the Mounties at Prince Edward Island patiently waited for the results.

Six weeks after the physical evidence arrived at NCI and was placed in separate locked refrigerators, we were ready to attack the jacket hair. Using a special cell culture laboratory originally designated for handling very dangerous viruses (e.g., Ebola or Lassa fever), Victor removed four hairs, each with a tiny piece of fleshy tissue on the roots. Marilyn cut each hair into its root and shaft and each was subjected to DNA extraction. They attempted to amplify one microsatellite using PCR for trace DNAs. One of the eight samples, a root, was successful. The other seven failed, probably due to complete decay of intact cat DNA. The DNA from the single successful hair root was then used to genotype the ten feline microsatellite markers selected as the most sensitive, robust, and unambiguous of the 250 microsatellites they had tried in preliminary testing.

The methods worked well and the composite genotype of ten different cat microsatellite markers was successfully determined for the jacket hair root. Seven of the microsatellite loci from the hair were heterozygous (they had two different-size alleles, one from each parent) and three were homozygous (they had only one size of allele since they inherited the same allele from both parents). Now they were ready to compare the jacket hair's genotype to Snowball's blood sample.

Marilyn had intentionally extracted the hairs first because that way there was no danger of contaminating the hair with Snowball's DNA, at least while the samples were in Frederick. Three weeks after the jacket hair genotyping, they removed Snowball's blood from the locked refrigerator and extracted his DNA. Then, they typed the same ten microsatellite loci for Snowball. In addition, the jacket hair sample was repeated alongside DNA from our NCI colony cats.

The results could not have been cleaner. All ten loci from Snowball matched the jacket hair genotype nearly perfectly: seven heterozygous loci, all the same, and three homozygous loci, the same, a total of seventeen allele matches.

There is an important reason why the match was nearly perfect, and not simply perfect. Microsatellite alleles are classified according to their migration distance on an electrophoretic gel (a Jello-like medium placed in an electric field that separates DNA fragments of different microsatellite allele sizes). The human forensic community got into some hot water in the early years of DNA fingerprinting by relying too heavily on the subjective opinion of DNA technicians (or their bosses) on whether or not crime scene samples matched evidence samples. By the time of our analysis two developments had improved things. First, the gel assays were automated using DNA genotyping machines that could measure the mobility (or size) of alleles extremely precisely. Second, the forensic community now agreed on objective criteria for pronouncing a match. Why was this second step necessary?

Even very precise genotyping machines, working on an identical allele, will produce slightly different measurements from gel to gel. For example, an allele with an actual size of 150 units (which is determined by the sum of repeat stutter letter nucleotides plus nonrepeat nucleotides in DNA that flank the microsatellite repeat) might be measured as, say, 150.02, 150.13, 149.91, 149.85, and 150.18 in five separate gel runs. The range of the actual measurements is called the "match window." The forensic community requires an explicit definition of and conformance to the match window criteria for microsatellite-based forensics, so we built one ad hoc for the cats. Fortunately, while mapping the feline genome, Marilyn and Victor had already genotyped a seventy-member lab cat pedigree for the ten microsatellite loci used to genotype Snowball. They measured the precise allele size and range for all eighty-seven alleles they found in the pedigree across the ten forensic microsatellite loci. So the allele size range and the match window was experimentally determined already. Credible allele match calls in our forensic case would have to fall within the match window for every single microsatellite allele at the ten loci.

By this standard, the jacket hair was a perfect match for Snowball. Did this mean that the hair definitely came from Snowball, connecting the blood-stained jacket to Beamish? Not quite.

The reason for caution here bears on how geneticists determine that a match, which we had, actually proves identity. The accepted approach is to ask this question: "What are the odds, or the statistical probability, that the hair genotype actually came from Snowball, and not from another cat on P.E.I. that by chance had the same DNA genotype?" The answer to that question depends on how commonly the ten-loci microsatellite genotype found in Snowball would occur by chance in the local P.E.I. cat population.

When we first found the DNA match, we had no idea of the frequency distribution of microsatellite allele genotypes in any domestic cat population, let alone on Prince Edward Island. The NRC DNA technology forensic report had said that the best way to get an accurate estimate of microsatellite allele frequencies for any case would be to have a population database of all genotypes from a sampling of individuals who lived in the neighborhood of the crime scene. We had no such database, and we needed to get one—fast.

I telephoned Constable Savoie and told him the good news—that we were able to genotype the jacket hairs and Snowball reliably. I told him that we might have a match but we had one last detail that we had to address before we could be certain. I explained that pronouncing an identity would require some knowledge of the levels and patterns of gene diversity in the local cats. Suppose the population was highly inbred, like the Gir forest lions or the Florida panthers. If so, Snowball's genotype could be so common as to be meaningless. I told Roger I wanted him to round us up some cats from around the crime scene on Prince Edward Island.

Savoie agreed and within a few weeks we had blood samples from nineteen cats, this time shipped by Federal Express. Marilyn and Victor ran the DNA genotypes of the nineteen cats for the same microsatellite markers to make a small DNA database. To our relief, the cat population sample had lots of genetic variation. Each microsatellite locus showed between five and ten different alleles, plenty to exclude any history of inbreeding. And nearly all the alleles observed in the jacket hair and Snowball were present in the population. This meant we could use the measured allele frequencies of the population to estimate (using basic statistics) the frequency of Snowball's genotype on Prince Edward Island. That frequency is the same as the chance that the matching hairs in the jacket did not come from Snowball, just what we were after.

Snowball's genotype frequency came out to be a vanishingly small number, 2.2×10^{-8}, or about 1 chance in 45 million. Since there were only a few thousand cats, maybe ten thousand at most in the entire province, it meant that the jacket hairs were indeed Snowball's. If Snowball had an identical twin or a clone, they would match as well, but cats don't produce identical twins and only years later would Texas A&M produce CC, the first cloned kitten.

Savoie was ecstatic. We were certain the hair came from Snowball and now we had the statistics and database, however small, to prove it.

Still, another potential mine field emerged from the human forensic community's concern about population substructure. If separate populations differ in microsatellite allele frequencies, strong biases play into computing an estimated genotype frequency, the likelihood of a chance match. This concern is particularly important among human racial groups that show very distinct microsatellite allele frequencies. We worried that there might be a possible problem in our computation of Snowball's expected population frequency from our small P.E.I. cat population database. Was the sample representative of other cat populations in Canada or the United States where the murderer's cat may have originated? We sampled the same microsatellites in a small group of cats from Maryland. This population's genetic structure was very similar to P.E.I. cats, with nearly identical allele frequency distributions. There were no obvious population substructure differences between cat population samples that were one thousand miles apart. Our concern about the population substructure vanished; we could not detect any, so the calculation was valid.

The Mounties in Prince Edward Island were still stymied by the lack of a body. Shirley Duguay had been missing for over six months. By then, Duguay's family and the police were convinced that she had been murdered.

On Saturday, May 3, 1995, Robert Nason, a local trout fisherman, discovered a shallow grave in the dense forest near North Enmore, ten miles from where Shirley's vehicle was abandoned. A team of forensic specialists descended on the remains of a young woman who matched Duguay's description.

The next morning, still awaiting a positive identification of the body but with knowledge of the DNA profile match of Snowball's hairs in the bloodied leather jacket, the Mounties arrested Douglas Leo Beamish and charged him with first-degree murder. Ten days later dental records confirmed the North Enmore body was indeed that of Shirley Duguay. Roger Savoie notified Marilyn that we were going to court and the evidence our genetics group had developed was absolutely critical.

When it became clear that the Snowball case was moving to trial, I briefed my boss, Dr. George Vande Woude, NCI division director. While emphasizing that the Snowball case had been a sideline to our everyday research work, which was directly related to health and cancer, I explained that this remarkable case could establish an important legal precedent for DNA profiling of animals. The Prince Edward Island crown prosecutor's office had paid for the expenses and laboratory reagents that were required for the exercise. However, we anticipated that Marilyn, Victor, and I would be subpoenaed for testimony. I asked for his advice and consent for our cooperation.

Vande Woude responded that he thought I was crazy but knew that before. He agreed that it seemed like an honorable venture and he understood why I had gotten involved. However, before he could approve our participation in the trial, he needed to run it by Dr. Richard Klausner, director of the National Cancer Institute.

The two directors passed my request on to the NCI ethics office. Weeks later Dr. Maureen Wilson, head of that office, told me how irregular my request was and that she was not comfortable approving it or advising Dr. Klausner to do so. She explained that federal employees are exempt from all subpoenas unless the U.S. government agrees to honor them, so I should simply ignore the subpoena.

I explained to Dr. Wilson that I *wanted* to honor the subpoena. After all, taxpayers had funded the developments of the cat microsatellites; they were still unpublished, and here was a way the people could immediately benefit from our research. She replied that that was not enough. I pressed that Dr. Klausner and Dr. Vande Woude seemed inclined toward approval barring any negative legal implications. But the conversation ended there. We had several similar telephone exchanges over the next few months. Beamish, meanwhile, had been in custody for four months and the Canadian Supreme Court was scheduling a preliminary hearing a month later, in August.

About a week before the hearing, we discussed the logjam once more and Dr. Wilson told me that the only way to get approval would be to seek permission from the Secretary of Health and Human Services, Donna Shalala. I said, fine, let's ask her. Wilson was not inclined to do this; it was too unusual and by implication not important enough.

I was at the end of my rope, so I changed my tack. My job is important to me and we clearly could not testify without NCI permission. So I asked Dr. Wilson to answer one question: When the Prince Edward Island and Canadian media contacted me to ask who decided I should refuse cooperation, a refusal that would lead to the release of an accused murderer in Prince Edward Island, should I answer Dr. Wilson, Dr. Vande Woude, Dr. Klausner, or Secretary Shalala? Where did the buck stop?

Victor, Marilyn, and I were en route to Prince Edward Island for the preliminary hearing when the NCI ethics office notified my secretary that permission to testify had been granted at last. The preliminary hearing was meant to review the evidence before a court justice to determine its admissibility. The hearing was important because DNA profiling had never been successfully introduced in a P.E.I. courtroom even for human materials, let alone for cats.

About six weeks before the hearing, we had learned that a Canadian court would consider forensic evidence only from a laboratory that was "certified" as having sufficient scientific expertise. My lab had never been certified for any such thing; I asked the Mounties what was required. They said our lab had to pass a blinded RCMP certification test.

Dr. Ron Fourney, head geneticist at the RCMP Central Forensic Laboratory, agreed to administer a proficiency test to us. Fourney sent a filter paper with eight spots of cat blood for us to analyze genetically. Victor and Marilyn extracted DNA and ran the ten cat microsatellite markers. Our analysis revealed three samples with a unique cat microsatellite genotype and two pairs with identical genotypes. The last sample gave very strange patterns. After a bit more detective work, that blood proved to be human. We sent our findings to Fourney, who affirmed we had gotten them all right. The official certification arrived by fax from Ottawa the evening we arrived in Summerside, Prince Edward Island, for the preliminary hearing of the Beamish murder trial.

The hearing was a minitrial, except there was no jury and a strictly enforced press blackout on the proceedings. Marilyn, Victor, and I had brought several thick ring binders packed with results, proficiency tests, controls, experiments, statistical analyses, a narrative on chain of evidence, and written reports of our findings. On the witness stand I explained to the judge the methods of DNA profiling with microsatellites, our match result, and our interpretations. I assured the court that we were experts in cat genetics and that our lab was recently certified by the RCMP to conduct such tests. Marilyn had prepared colorful cardboard charts with illustrations of microsatellite profiles, the allele frequencies of the cats in the P.E.I. database, and computations of Snowball's genotype frequency. The questions and cross-examination were not as probing or as aggressive as we had feared. Our testimony was successful and the trial was scheduled for May 21, 1996. We now had nine months to prepare for our testimony.

In the interval between Shirley's disappearance and the preliminary hearing, the whole world got a lesson on DNA profiling and courtroom drama from comprehensive TV coverage of the sensational O.J. Simpson murder trial. Simpson stood accused of brutally murdering his ex-wife Nicole Brown and her friend Ronald Goldman at her home outside of Los Angeles. Human DNA profiles were collected and analyzed from five different places: 1) the crime scene at Nicole's Brentwood condominium; 2) O.J.'s Ford Bronco; 3) O.J.'s house; 4) socks found in O.J.'s bedroom; and 5) a glove outside O.J.'s house. Forty-five blood specimen DNA profiles were conclusively matched to either Simpson at the crime scene or to Brown and Goldman at Simpson's house and in his Ford Bronco. These genotypes were resolved with no other unexplained person's genotypes at any place. Undaunted by the genetic evidence, the legal defense team skillfully assaulted the chain of evidence, the motives and integrity of the LAPD officers, and the reliability of dubious and confusing scientific uncertainties. The jury discounted the compelling DNA evidence and found Simpson not guilty. He was acquitted on October 3, 1995, exactly one year after Shirley Ann Duguay went missing from her home in Prince Edward Island.

The Simpson case was not the first, nor probably the last, time that genetic typing went down the drain, or should I say over the jury's head, in a high-profile case. In a World War II–era version of the "trial of the century," accused comic actor Charlie Chaplin was excluded as being father of plaintiff Joan Barry's daughter by genetic typing of red blood cells. Ms. Barry had type A blood, the daughter had type AB, and Chaplin had type O. It is genetically impossible for a couple with type A and type O blood to produce a type AB child; the child's B allele type must have come from another father.

Joan Barry's flamboyant lawyer ignored the mysterious science and simply held the newborn baby before the jury box, imploring them, "Doesn't this cute little girl look a lot like Charlie?" Chaplin was ordered to pay child support. The verdict destroyed his career and his life.

I was inwardly terrified, but determined to avoid the perils of legal genetic confrontations that had come before. Our data were clean and crisp and it was my job to make the evidence digestible to a jury of twelve Prince Edward Island citizens.

The trial went on for eight weeks; it involved 160 exhibits and over a hundred witness testimonies. Following British tradition, the lawyers and judge wore black robes, albeit without the wigs. Chief Justice David Jenkins, the presiding judge, was a bright and stern arbitrator who had little patience with legal wrangling or stalling. That was a good sign, I thought. David O'Brien, the crown prosecutor, the Canadian equivalent of a district attorney, was a perspicacious, easily flustered, and often disheveled perfectionist. His meek academic manner at one stage prompted Justice Jenkins to ask O'Brien to quit mumbling and explain himself more clearly. O'Brien was tireless in his attention to detail and remarkably prepared for the uncertainties of a capital murder case with no eyewitnesses and rather complex scientific evidence as his basis for prosecution.

The forensic pathologist who examined Shirley's body testified that her jaw was broken in three places, as was her nose, and that her Adam's apple was crushed. Her hands were tied behind her back with a clothesline when her body was uncovered inches below the surface in the woods near North Enmore. She had been beaten to death, mercilessly.

Crown Prosecutor O'Brien produced a letter, apparently written in blood, to Shirley Duguay from Doug Beamish weeks before her murder in which he threatened her. He wrote that if she would not take him back once again he would have no reason to live. He might as well kill her, the children, and himself.

O'Brien then entered a photograph of Beamish wearing a brown leather jacket, which was to my eyes identical to the brown leather jacket found spattered with Shirley's blood in the woods near Tyne Valley. A friend of Doug Beamish's testified that he had taken the photograph of Beamish in the jacket a few days before the murder. The leather jacket had been hung behind the judge for all to see throughout the trial.

Another witness related a conversation in which Beamish, upset about Shirley's new boyfriend, sputtered, "That woman should be shot . . . and pissed on!"

When Marilyn, Victor, and I arrived at the courthouse in Summerside, P.E.I., the trial had already been under way for three weeks. Crown Prosecutor O'Brien met with us the night before we were to testify and carefully reviewed our strategy until nearly midnight.

The next day the courtroom was half empty. It was scattered with a few townspeople, but the dominant presence was the Duguay family—sisters, parents, cousins—on one side, with the defendant and the Beamish clan on the other. The territorial separation of the two families was a source of palpable tension. In my imagination the Beamish group appeared angry, even menacing. Marilyn became agitated and unsettled during the proceedings. She moved away from the Beamish family to avoid their spooky glances. At one point she asked Victor to check under our rental car hood for possible explosive devices. The growing anxiety made Victor lose his appetite for the entire week.

To qualify as expert witnesses, Marilyn, Victor, and I had to present our credentials and be cross-examined with the jury excused. Crown Prosecutor O'Brien went through my curriculum vitae, singing my scientific praises for about thirty minutes. The defense counsel, John MacDougal, was youngish, handsome, and articulate, and had a crisp sense of humor. As he approached me, I suspected that the jury would like him.

MacDougal challenged my credentials as an expert witness aggressively. He questioned my eight adjunct university appointments and several hundred publications. Were the university posts honorary and meaningless? How could anybody write so many papers? He counted the titles among my scientific publications and announced that only 10% were about cats.

I countered that universities conferred degrees on my graduate students and that many articles on cats did not list cats in the title. MacDougal focused on one paper entitled "A Canine Distemper Virus Outbreak in Serengeti Lions." "Doesn't 'canine' mean the study is about dogs, not cats?" he demanded. I said he was correct that "canine" refers to dogs, but the title is about a dog virus that had leaped into lions, and lions are cats. After forty-five minutes of such berating, he gave up and the judge qualified me. Interestingly, it took less than five minutes of cross-examination to qualify Marilyn and one minute to qualify Victor.

Roger Savoie had testified earlier about collecting the hairs, the blood, and the population samples. He described his personal transport of jacket hairs and Snowball's blood to Frederick to preserve the chain of evidence. Marilyn and Victor detailed the careful procedures they had used to extract DNA, determine genotypes, avoid contamination, and control their experiments. The jury listened carefully.

My challenge was to explain to the jury how the microsatellites from the jacket hair represented a unique genetic signature, something like a name, rank, and serial number for any individual. Then, using Marilyn's posters, I walked them through the microsatellite genotyping of the jacket hairs and Snowball's blood sample. To be thorough I also explained the match window standards to show we were aware of some of the historic criticisms of human DNA profiling and how we avoided them. Finally, I stated that there was a clear match between the jacket hair and Snowball's blood—all ten microsatellite loci, seventeen alleles.

At the end of the day Prosecutor O'Brien left the jury with a few rhetorical questions. What does a match mean legally? What is the probability that the hairs really came from Snowball? Put another way, what is the chance that a cat other than Snowball in Prince

Edward Island has the same genetic type? And can we conclude that Snowball's hair had found its way to the lining of the jacket hanging over the judge's bench?

As the court adjourned, my partners and I worried among ourselves. How do we explain the probability computation, the meaning of statistical likelihood to common citizens—to construction workers, housewives, and TV cable repairmen? We conjured up a colorful analogy to Canadian's national pastime, ice hockey. The NHL Stanley Cup playoffs were under way and the whole town was talking about it. Don MacLean, a Summerside native, had taken the Florida Panthers (a hockey team from Tampa, not the puma subspecies) to the championship in his first year as head coach.

Although I did play hockey for a few years as a teenager, none of us were avid hockey fans, so I wanted to be sure of one thing. At the Burger King where we breakfasted the next morning, I asked the fellow behind the ovens flipping omelets how many players make up the first string of a hockey team. Before he could answer, the young woman who had served us our egg croissants blurted out, "Why didn't you ask me that question?" I fidgeted, quite embarrassed by my gender bias, and then meekly asked her the same question.

"I have no idea," she shot back. Then, the cook answered, "Six players, including the goalie!" I made a quick calculation on a napkin and we went to court.

On the stand I asked the jury to accept that estimating the statistical probability of the jacket hairs belonging to a cat *other than Snowball* was the same as the frequency of Snowball's (and the cat hair's) composite microsatellite genotype in P.E.I.

In statistics, a rule you must remember is that if you know the probability of some event, A, and also of a second unrelated or independent event, B, the chance that both A and B happen is known simply by multiplying the probability of A times the probability of B. For example, if I throw two dice, the chance of throwing snake eyes, two ones, equals 1/6 times 1/6, or 1 in 36.

So for a composite ten-locus microsatellite genotype, the frequency in a population is determined by multiplying the population frequency of the first microsatellite locus genotype times the second locus frequency times the third, all the way to the tenth microsatellite locus. DNA forensic geneticists call this multiplication scheme the "product rule." Fortunately for us, six months before the Beamish trial, the U.S. NRC report on DNA technologies in forensic science had come forward with a strong endorsement of the product rule in cases like this.

We had measured the frequency of single microsatellite locus genotypes in our ad hoc P.E.I. database of nineteen cats, so we could multiply the values to get Snowball's estimated genotype frequency. That product computed to a very small number, about 1 in 45 million. This meant our matching of microsatellite genotype from jacket hairs to Snowball's blood was compelling evidence they came from the same cat, our hero Snowball.

I cautioned that the computation was theoretical, but the forensic scientific community had endorsed it as rather reliable. To help clarify how the calculation worked, I asked the jury to imagine a convention at the local Prince William Hotel, involving ten teams from the National Hockey League. Suppose each team assembles their first string, six uniformed players, in ten separate rooms. In the first room are six players from the Montreal Canadiens, in the next, in the first string of Edmonton Oilers, in the third, the starting six from the Vancouver Canucks, and so on for all ten team rooms.

Now suppose I open the door in the first room and blindly toss a hockey puck and see who catches it first. The probability that it would hit the Montreal goalie in the first room would be 1 in 6. Now suppose I throw the puck in the second room. The chance to hit a goalie there would also be 1 in 6. But the chance that two goalies are hit in the first two rooms is 1/6 times 1/6, or 1/36. Now if I throw the puck in all ten rooms, the chance it hits a goalie in every single team room is $1/6 \times 1/6 \times 1/6$ ten times over, or 1 in 60,466,176. This unthinkably low value reflects how unlikely it is to expect to randomly hit 10 goalies in 10 rooms by chance.

The 1 in 45 million estimate for Snowball's genotype had a similar meaning. An extensive DNA genotype match said that the jacket hairs belonged to Snowball. The chance that the hairs did not come from Snowball is such an extremely rare probability that it was as close to impossible as we could imagine. The simplest explanation by far for all this statistical analysis is that the hair provided to us in that evidence canister and Snowball's blood sample both came from the same cat.

Defense Attorney MacDougal was not pleased. He began his cross-examination with a question.

"Dr. O'Brien, are you familiar with the O.J. Simpson trial?"

"Somewhat, but perhaps not as familiar as you or others in this courtroom," I responded.

"Do you recall the well-known quote: 'If the glove doesn't fit, you must acquit'?"

"Yes."

"Did the Mounties by any chance say to you, 'Without the cat the case falls flat'?" he asked.

"No, I don't think they did."

I could imagine this sound bite as the next day's newspaper headline. MacDougal went on to criticize and to attempt to obfuscate the chain of evidence, the genotype certainty, and our temerity for entering DNA profiling from a house cat in a murder trial.

At one stage MacDougal approached me and theatrically pulled a piece of my own hair off my sports jacket. "Now, Dr. O'Brien," he mused, "I seem to have found a white hair in the wrong place here. Is it possible that a similar incident might have occurred above your vials of the jacket hairs?"

Taken aback, I swallowed hard and answered sheepishly, "No, I do not think so, for two reasons. First, handling of the samples was all done by Mr. Victor David and Dr. Marilyn Raymond in biohazard containment cell culture hoods as I occasionally watched from a safe distance. Second, when this analysis was under way, I was struggling with the implications of my approaching middle age. My hair at the time was dyed dark brown!" Victor lowered his eyes. Marilyn smiled and nodded in affirmation of the story to the noticeably amused jurors. I guess that's why lawyers insist that one should never ask a question without knowing the answer.

Nearing the end of his cross-examination, MacDougal sputtered out, "I suggest to you, Dr. O'Brien, that this one in forty-five million is merely a piece of theoretical bullshit!" The judge reprimanded him for the outburst.

Startled by his disdain and abject ignorance, I returned a parting jab at MacDougal. "Counselor," I responded, "I respectfully disagree and I would add that I am sorry that you do not seem to get it. But I do think the jury does get it."

The trial went on for a few weeks after our testimony. In the end, the defense called no witnesses, arguing to the court that the Crown had failed to prove its case.

The jury deliberated for two days and delivered a unanimous verdict of guilty of murder in the second degree (unpremeditated). Two months later Beamish was sentenced to eighteen years in prison without the possibility of parole. At the sentencing, Justice David Jenkins told Beamish that he displayed "a callous disregard for human life. The murder was brutal; the circumstances were horrific."

A series of candlelight marches in quiet protest of unanswered domestic violence occurred in Charlottetown, P.E.I.'s capital city, that summer. Nelson Beamish, brother of Douglas and married to Shirley Duguay's sister, commented on the conviction, "Shirley had to die to prove my brother was dangerous. Good riddance." Two years later, torn by the turmoil of losing his sister-in-law and his brother, Nelson Beamish committed suicide.

The conviction of Douglas Leo Beamish that July was not big news outside of eastern Canada. Cameras had been banned from the courtroom and only a few vigilant Canadian reporters covered the entire proceeding. Nonetheless, the trial did set an important international legal precedent: the introduction of automated microsatellite genotyping—DNA profiling—of pet animal hairs in a capital homicide case. In 1997, Marilyn, Victor, and I published a brief summary of the case in *Nature*, which was picked up by the national media.

Once the legal and forensic communities learned of the case, the headline writers had a field day. "Fur-ensic Evidence" rang one in *Scientific American*. The HBO-TV series *Autopsy* called Snowball the "Purr-fect Witness," while the *National Enquirer* reported a "CAT-astrophe for Criminals." Associated Press heralded an era of DNA "Paw-Prints."

As the story became widely publicized, I started getting calls from criminal investigators. Animal hairs were turning up at a murder, a rape, a bombing, and other serious crimes. Soon we were inundated with requests from detectives to help nail their suspects.

I had to decline these many requests, but by then private forensic laboratories were taking such evidence a little more seriously, gearing up for cat and dog DNA profiling. An estimated sixty-five million pet cats live in the United States. That number translates to a cat in every third or fourth household, including the households of criminals and their victims.

Criminals almost invariably leave biological materials at crime scenes—on weapons, clothing, vehicles, telephones, condoms, utensils, even on doorknobs and light switches. If just a quarter of the twenty thousand murders per year in the United States were committed by cat owners, a significant fraction would leave their pet's DNA at a crime scene.

In 1999 the U.S. Department of Justice awarded our lab a grant to develop a microsatellite-based population genetic database for the forty recognized breeds of pet cats. When the database becomes available, there will be no need to assemble an ad hoc population database of the cats residing in the crime scene neighborhood. The new feline database would provide an important addition to the resources available in criminal forensic investigations.

It is now clear that an immediate human benefit of the new genomic technologies lies in criminal justice. Although DNA profiling will implicate guilty suspects, more frequently it will exonerate those who are falsely accused. One early FBI estimate projected that DNA profiling has directed investigation away from the prime suspects in capital cases as often as 30% of the time. Philip Reilly, an author of the 1992 NRC DNA forensics report, speculated that at least 5% of convicted rapists are wrongly imprisoned. The New York City–based Innocence Project, led by lawyers Peter Neufeld and Barry Scheck, pioneers of DNA profile forensics, has exonerated and

gained release for over fifty convicted felons, including several awaiting execution. DNA profiling of people has become the gold standard
of forensic evidence. When properly collected it is very difficult to
assail.

Cumulative DNA databases of convicts have been established in
the United States to allow for what Reilly calls a "cold hit": matching
DNA at a crime scene to the genotype of a former detainee. There
are literally tens of thousands of "rape kits"—biospecimens from past
crimes stored in police lockers. Can these materials be added as
microsatellite genotypes to a worldwide genotype database of people
linked to crime scenes? Although the need for a DNA-based
approach is bolstered by the high rate of recidivism among felons
(estimated at 40–60%), ethical and constitutional concerns about vio-
lation of the Fourth Amendment right to protection from unreason-
able search and seizure merit closer examination. So far court
challenges attacking cold hit DNA databases of crimes have not been
successful, but the expansion of broader DNA-profile databases
makes some civil libertarians shiver.

As I write these words, Douglas Leo Beamish is serving his sentence and has appealed his conviction. Snowball still lives with his
family, although Beamish's mother revealed on a TV interview that
the cat has never been the same since the Mounties came. Marilyn
and Victor published the first microsatellite genetic map of the cat in
1999. We are trying to persuade the NIH to fund a full DNA
sequence of the cat genome. Roger Savoie received the high honor of
being named the Mountie of the Year in 1997 for his persistence and
dedication.

Twelve

Genetic Guardians

OF COURSE I HAD BEEN HERE BEFORE—PLENTY OF TIMES—but it never had seemed quite so still. Growing up in Bethesda, I made trips to the Capitol Mall in downtown Washington, D.C., so often as to feel blasé about them. An exception was that fierce blizzardy morning in January 1961 when my high school pals and I made our way to the Capitol to hear young President John Kennedy utter his timeless inaugural plea, "Ask not what your country can do for you, ask what you can do for your country."

Strolling across the Mall on Columbus Day, October 12, 1996, my otherwise loquacious wife, Diane, and my two equally gregarious daughters were uncharacteristically somber. We were surrounded by the 40,000 colorfully embroidered patterned fabrics that comprised the AIDS quilt. Each section was meticulously stitched in memory of a fallen victim; the endless rows gave lasting testimony to the enormity of this modern scourge. The quilt's expansiveness took our breath away; the sections stretched down the Mall between the dozen Smithsonian museums, extending to cover the green from the Washington Monument to the steps of the Capitol building. It would be the last public display of the full AIDS quilt. With so many victims, the quilt had grown too large.

Families of AIDS victims wandered quietly, reverently, among the brilliantly blazoned emblems, emotion-charged and teary. I had watched my only brother, Danny, succumb to the disease in 1994, a familiar casualty of the "gay plague," as it was called for a brief moment in early 1980s San Francisco. There was no quilt square here for Danny. Diane had asked a matronly seamstress at my children's village church to prepare one in his memory, but her unabashed revulsion at the "sinful" disease prompted her refusal.

Nor was there a quilt square for 90% of the 350,000 young Americans who had died with AIDS by then. There were only the vast symbolism and the saddened survivors that day on the Mall. We mourned together a horror that had touched each of us.

My daughters, then aged sixteen and eighteen, had never known a time that was free from fear of AIDS. They had watched their favorite uncle dwindle away in a culture of alternating judgment and compassion, victim of a plague that respected no boundaries, nationality, ethnicity, age, intelligence, or social status.

Walking among the poignant Technicolor images, I recalled the celebrity casualties: Rock Hudson, Ryan White, Arthur Ashe, Elisabeth Glaser, Rudolf Nureyev, Liberace. We barely took notice as Elizabeth Taylor took to the podium set up on one end of the Mall. The movie star turned social advocate spoke of compassion, of hope in education, in prevention, and in basic AIDS research. I appreciated her support.

Only a few weeks before, our laboratory team had announced the discovery of $CCR5\text{-}\Delta32$, the first human gene variant that conferred complete resistance to HIV infection on its carriers. Such genes provide hope for a new kind of immunity to a deadly scourge for which, so far, there is no real cure or any effective vaccine. The trail leading to the discovery of $CCR5\text{-}\Delta32$ and other elusive AIDS resistance genes (we also call them "restriction genes") was punctuated with determination, patience, endurance, preparedness, and serendipity. Our search began in the early 1980s when the first signs of acquired immunodeficiency syndrome—AIDS—appeared in the Western world.

On June 5, 1981, Los Angeles immunologist Dr. Michael Gottlieb published a short report in a weekly Centers for Disease Control medical journal describing five cases of debilitating *Pneumocystis carinii* pneumonia among young, actively gay men. That agent is a ubiquitous microbial Protozoa that rarely causes disease except in newborns in intensive care or in people with compromised immune responses such as cancer patients being treated with immune-suppressive drugs. Gottlieb's patients also had unusual mouth sores caused by *Candida albicans* fungi. Three of his patients had an inexplicable collapse in their immune system. Shortly thereafter, a heretofore very rare purple-blotch skin cancer, Kaposi's sarcoma, began to appear among young gay men in New York, San Francisco, and Los Angeles, also accompanied by extreme immune suppression.

What would briefly be called the "Gay Related Immunodeficiency Disease" (GRID) grew in its incidence by the month, and soon it afflicted surgery patients who had received blood transfusions. Then hemophiliacs, whose hereditary syndrome is treated by regular volunteer blood transfusions, started to exhibit immune suppression and hypersensitivity to *Pneumocystis,* to cytomegalovirus-mediated blindness, and to a variety of secondary infections. The clustering of immune disease in gay men, transfusion recipients, and hemophiliacs was a sure sign of an infectious agent, one that had contaminated the blood banks.

By 1984, the cause of AIDS was shown to be a lentivirus (a slow-growing retrovirus related to scrapie, a neurological-disease-causing virus originally discovered in sheep and horses) called "human immunodeficiency virus" (HIV). That discovery is officially credited jointly to Luc Montagnier of Institut Pasteur in Paris and Robert Gallo of the NIH, and later affirmed by Jay Levy at the University of San Francisco. (In reality, the discovery was tangled in fierce rivalry and accusations of scientific marauding, deception, chauvinism, and fraud that rattled the halls of science and Congress.) Once the virus had been identified, first by Montagnier, a blood test was quickly developed to detect antibodies against HIV in patients' blood. Gallo developed a blood test for HIV that was licensed in 1984, an event that rapidly led to the cleansing of Western blood supplies from the deadly virus.

HIV/AIDS has since spread across the globe with a furor unmatched since the Black Death. Nearly every grim epidemiology prediction for the course of HIV spread has been met or exceeded. The statistics are numbing. In North America alone, some 448,000 young men and women have died of AIDS. Over a million Americans have become infected with HIV. But the American toll represents only the tip of a global iceberg.

As I write, 22 million people have died from AIDS worldwide; and 42 million men, women, and children, more than the population of Australia, are now infected with HIV. During the year 2002, 3 million people died of AIDS and 5.5 million people became newly infected. Worst hit is the African continent, accounting for 70% of new infections. Several African countries have an HIV infection incidence of over 30%. Life expectancy in Botswana, home to a horrific 40% HIV prevalence rate, has dropped from sixty-one to thirty-nine years. In the next decade, sub-Saharan Africa will be home to 40 million AIDS orphans.

Left untreated, half of HIV-infected individuals die of AIDS within ten years of infection. With a few exceptions, the rest will die over the next decade. There is no effective vaccine and no cure, although a powerful anti-HIV triple drug therapy can postpone AIDS symptoms for several years. HIV infection is more deadly than a bullet to the head; it just takes longer. Moreover, the incidence of infected people is doubling every few years. One 1998 United Nations forecast projected world population growth downward from 9.4 billion to 8.9 billion in the year 2050, a reduction of 500 million people largely due to expected increases in AIDS mortality. *Newsweek,* on the twentieth anniversary of the first AIDS cases, wrote, "AIDS is twenty years young. The worst plague in modern history is far from done with us."

When I first learned of AIDS in the early 1980s, I was looking for a serious human epidemic to study. I had become persuaded by the Lake Casitas mouse and feline virus examples that intrinsic genetic variants play a large role in the response of a population to a fatal infectious disease. It was time to search for such restriction genes in humans. Once discovered, they would help fill in the gap in our knowledge of gene-mediated immune defense. Perhaps some clues from natural genetic defenses could even be translated into new treatments and, one day, possibly a cure.

I was never very confident that medical scientists could guess at all possible therapeutic approaches for deadly diseases; I was more optimistic that countless generations of trial, error, and natural selection had come up with innovative genetic solutions for historic deadly diseases. So I set out to find human genes that would influence the outcome of exposure to or infection by HIV. It was a hunch with little firm basis and no guarantee of success, but a potentially huge payoff.

The quest would not be an easy one. First, there was no tried and proven recipe for finding disease-resistant genes in humans; we had to learn from experimental trials and our own missteps. Second, human genetics was in its infancy. In the early 1980s fewer than 1,000 genes (of the 35,000 in our genome) had been mapped or even described. Third, even as AIDS developed into a full-blown epidemic, there was no evidence that human genes had a role in the disease. Indeed, more frequently than I care to recall, I encountered harsh skepticism over my hopeful and very pricey scheme. Some critics dubbed our project a scientific "fishing expedition."

In the beginning, I was a bit uncertain myself and sought advice from my friends. Several human geneticists supported the concept of finding AIDS restriction genes, but they themselves had little interest in personal involvement. In truth, they were worried about exposing their staff and students to anything as deadly as HIV. This was a very scary virus. Knowledgeable virologists I contacted also thought the project was a bold and provocative idea, but not being geneticists themselves, they had no idea how to harness the rapidly developing tools of human molecular genetics to find such genes. My own background in feline retrovirology, population genetics, and human gene mapping seemed a nice fit, if only I could get the cash needed to proceed. I decided to push full steam ahead.

First, I persuaded my boss, Dr. Richard Adamson, division director at NCI, to bankroll the hunt. He supplemented my budget with close to a million dollars. Then I hit the road, hoping to enlist AIDS epidemiologists to join in a collaboration. Epidemiologists study the differences in how people succumb to or resist epidemics. Working in public health settings at universities and local municipalities, they track large groups of patients in tedious detail, cataloging the social, cultural, and environmental variables that could account for different outcomes in an afflicted population.

As in any epidemic, the impact of HIV exposure varied from patient to patient. For example, of the 12,000 American hemophiliacs who had received HIV-contaminated blood products between 1978 and 1984, before HIV screening of blood began, around 85% became infected, but 15% did not. Several who became infected had brothers who did not, even though the two brothers received infusions from the same contaminated blood injected by the same doctor. Could it be that the lucky brother's genetic complement played a role in resisting infection?

Another unpredictable aspect of the AIDS epidemic involves how long it takes any HIV-infected person to progress to AIDS or to die from immune collapse. The median time from HIV infection to AIDS-defining diseases is about ten years. However, some people succumb in less than twelve months, while others, a small percentage, avoid all AIDS symptoms for twenty years or longer. Was that a genetic influence?

Also puzzling were the types of AIDS-defining diseases that would develop. Some AIDS patients would get *Pneumocystis* pneumonia, and others would develop Kaposi's sarcoma. The myriad of other AIDS-defining diseases include tuberculosis, lymphoma, neuropathy, dementia, blindness due to cytomegalovirus infection, candidiasis yeast infections, and liver failure. Did different human gene variants predispose or resist these different clinical outcomes? No one knew.

In the coming years, I contacted scores of epidemiologists involved in developing longitudinal AIDS cohorts—patients in different groups at risk for HIV infections, for example, gay men, hemophiliacs, intravenous drug users, and HIV-positive mothers and their infants. I asked them to provide me with a single blood sample from each patient they saw. I turned the blood over to Cheryl Winkler in our laboratory (having progressed from cheetah skin grafts to AIDS research), who made immortal B-lymphocyte cell lines using a remarkable virus called Epstein-Barr virus, or EBV. This virus is well known as the cause of teenage mononucleosis, and in rare cases, nasopharyngeal carcinoma and Burkitt's lymphoma. Less well known is the ability of EBV to "transform" human white blood cells into an immortal cell line in the laboratory, thereby providing us with a renewable and limitless supply of patients' DNA.

In those early years I enlisted some twenty AIDS cohorts and collected over ten thousand patient blood specimens. Each sample was carefully transformed; the cells were cultivated and DNA was extracted. With every year's passing, the extensive clinical details of each volunteer's progression to AIDS expanded the computer databases of my epidemiological collaborators. Soon we had enough information to begin connecting the clinical data to the genes of each patient. This meant integrating one growing discipline, AIDS epidemiology, with another, human molecular genetics.

Years before the Human Genome Project completed its full draft sequence, the extensive human DNA research had uncovered considerable genetic variation among people and peoples. Relatively common genetic DNA variants appear every twelve hundred nucleotide letters in our genome, providing five to ten million genetic variants, any one of which might influence how a victim would respond to HIV. I believed that by a comprehensive population-based genetic screening of the cohorts, we could link certain human gene variants to AIDS resistance. The approach we took was simple, but it was soundly predicated on a generation of population genetic theory and practice.

My epidemiologist colleagues and I divided each AIDS cohort into different disease categories—groups of HIV-exposed patients who shared similar health or clinical outcomes. For example, to look for a gene that would block HIV infection, we gathered individuals who were infected with HIV (identified as being HIV-antibody-positive in a blood test) and a second group that never became infected even though they had been exposed to HIV. Exposure was assessed based on an admitted history of risky sexual behavior among gay men, documented receipt of known contaminated blood-clotting factor among the hemophiliacs, or needle sharing among intravenous drug users in a city slum with a high incidence of HIV infection.

We also compared HIV-infected patients who progressed rapidly to AIDS-defining diseases with patients who avoided AIDS for a long time; by 1990 we had hundreds of slow progressors or "long-term survivors" who remained healthy for ten or more years after HIV infection. Finally, we partitioned AIDS patients as to their specific AIDS-defining disease: *Pneumocystis carinii* pneumonia, Kaposi's sarcoma, lymphoma, neurological disease, and other maladies.

In the mid-1980s I hired Dr. Michael Dean, a superb young molecular biologist, to develop the human genetic technology for this project. Mike's job was to discover and assess variants in all the human genes that AIDS researchers could tie to HIV infection and AIDS disease.

The horror of AIDS had stimulated a flurry of basic research investigations focused on the mechanisms by which HIV enters the body, kills cells, and destroys the immune system. That search had identified several human cell proteins that were usurped from their normal functioning to "collaborate" with the invading virus in its march to destroy a person's immune system. The genes that encode these cooperating cellular proteins immediately qualified as candidate genes for specifying AIDS resistance. Now we needed to discover their naturally occurring gene variants in hopes that some would influence the outcome of an encounter with HIV.

So Mike Dean began combing the AIDS research literature for candidates, then sequencing the genes in search of common gene variants. Over the years, he assembled several hundred gene variants to test in our AIDS cohort populations. For each new gene variant Mike discovered, he measured allele frequencies among different AIDS disease categories. Then we compared the genotype frequency in each category as well. (Genotype is the sum of two alleles, one from each parent, that a person has at each gene locus.)

If we discovered a clear difference in allele or genotype frequencies (or both) between two compared disease category groups, we would be on to something. It might mean that the gene variant in excess in one group was predisposing individuals to fall in that group and not the alternative category. For example, if a candidate gene had two allelic variants, A and B, and A was disproportionately high in HIV-infected and low in exposed uninfected individuals, then we would suspect that carriers of A were more sensitive to HIV infection, while those with B were resistant. However, if the allele frequencies of A and B were the same in both disease categories, we would conclude that the variants had no influence on HIV infection or transmission.

For years, the clinic doctors and nurses, our epidemiologic collaborators, Cheryl Winkler, Mike Dean, and I continued to add more patients, more genes, more genetic variants, and more sophisticated computer programs to search for aberrant allele or genotype distributions. By the mid-1990s, we had screened thousands of patients for over 100 candidate gene variants and another 250 additional DNA variants outside but nearby other coding genes distributed across the human chromosomes.

Every so often, we thought we spotted a genetic difference among groups, but they all evaporated under closer inspection. Meanwhile, we would monitor the many new research advances that were reported in the AIDS literature, searching for new genes to test. Finally, twelve years after we had begun what was becoming a tedious, very expensive, and thus far disappointing search, there appeared a glimmer of hope. The year was 1996.

"The End of AIDS?" proclaimed the *Newsweek* headline in December of that year. *Time* magazine followed by naming AIDS researcher David Ho their Man of the Year, and predicted that 1996 would be remembered as the turning point of the AIDS epidemic. That year AIDS research had witnessed two important and promising advances that tiptoed toward the unraveling of AIDS progression and perhaps defending against it.

The first advance was the dazzling success of new, powerful anti-HIV drugs. The announcements came from the Vancouver International AIDS Congress in January. A triple combination of potent new pharmaceuticals blocked two HIV-specified enzymes, protease and reverse transcriptase, and could reduce HIV to levels undetectable in serum, flat zero in AIDS patients. Widespread use of the combination drug treatment over the next several years would slow AIDS progression and cut AIDS mortality in Western countries dramatically. The treatment led to a dramatic 60% drop in the U.S. death rate, from 50,610 AIDS mortalities in 1995 to 16,273 in 1999.

Unfortunately, these powerful drugs were not able to wipe out the virus from HIV-infected patients completely—suspension of drug treatment even after years of negative blood tests invariably leads to a rebound of HIV. It seems HIV hides in still unidentified tissue reservoirs for years throughout treatment, poised to pop back when its poison disappears. Also, the drugs do not work for everyone; 40% of those who begin the complicated and often debilitating regimen fail. The anti-HIV drugs are frequently toxic, nauseating, and have several nasty side effects, hardly a perfect drug for lifelong therapy. Equally relevant, the therapies are very expensive, $10,000 to $20,000 per year, making them completely unaffordable in the developing world where they are most needed.

The second breakthrough in 1996 improved considerably our understanding of how HIV escapes from and destroys the immune defenses so effectively. The new insight grew from incremental experimental advances by many specialists—virologists, molecular biologists, X-ray crystallographers, immunologists—all probing to unravel the strategy of the advancing AIDS virus. I shall not explain the intricate details of how these discoveries transpired, as they are rather technical and well described in scientific articles referenced for this chapter at the end of the book. Instead, I will summarize briefly how we now believe HIV accomplishes its deadly damage.

A successful HIV infection usually involves delivery of an HIV-infected cell directly into the bloodstream of the recipient or through sexual contact. Free virus, such as occurs in saliva or even semen, is not highly infectious. Unless packed in living blood cells, the virus will infect only when delivered in whopping doses. This does not happen very often, but tragically it did occur in some HIV research laboratories, when a few skilled technicians became infected accidentally through open sores on their skin after handling concentrated laboratory strains of HIV.

Once HIV gets a foothold on its victims, it seeks out a variety of tissue cell compartments in which the virus can replicate. These can be lymph nodes, neural tissue, or epithelium in the gut or vagina, spleen, testes, kidneys, and other organs. HIV's principal factory is a grouping of three kinds of lymphocytes (white blood cells): macrophages, monocytes, and T-lymphocytes, all of which carry a cell surface protein called CD4. CD4-bearing T-lymphocytes are specialized immune cells equipped to seek out foreign viruses like HIV and to destroy them. In a sinister fashion HIV infects and demolishes the very cells that are meant to dispatch invading viruses.

HIV enters various cells by co-opting two receptor proteins on the cell surface. The CD4 molecule acts as a docking station, which hooks the HIV surface envelope protein and leaves the virus hanging atop the cell. Then a huge cell surface protein receptor named CCR5, which floats around the fluidlike cell surface membrane, meanders into contact with the CD4-snagged HIV. The CCR5-HIV interaction stimulates the cell membrane to dissolve enough to engulf the virus into the cell.

CCR5 normally serves as one of twenty cell surface chemokine receptors. Chemokines are short proteins, 100–125 amino acids long, that are released by tissues damaged by abrasions, bruises, or even infections. The chemokines attach to receptors like CCR5 on the cell surface of lymphocytes patrolling in the bloodstream, thus alerting the lymphocyte to alter its course toward the injury or infection.

The discovery that HIV utilizes both CD4 and CCR5 to enter the cells was rather important, culminating in historic papers from five separate groups published within a week of each other in *Science, Nature,* and *Cell* in late June 1996. The research groups were led by Drs. Bob Doms, University of Pennsylvania; Joe Sodrowski from Dana Farber in Boston; Ed Berger of NIH; and two groups from David Ho's Aaron Diamond Research Institute in New York, one headed by Richard Koup and John Moore, the second by Ned Landou and Dan Littman.

John Moore conjured up a useful image to envision the process of

HIV infection. Imagine HIV as an airborne blimp full of people en route to the subway tunnels below New York City. The Empire State Building represents the CD4 molecule sticking up from the cell, that is, Manhattan. The blimp hooks to the building steeple and waves in the breeze. Eventually a large cable car elevator, the CCR5 molecule, adjoins the dirigible, off-loads the human cargo, and delivers the passengers down to the subway station. Manhattan—the CD4/CCR5-bearing T-cell—has become infected with HIV.

As HIV sets up housekeeping in the macrophages, monocytes, and T-cells, it somehow remains sequestered or protected against immune clearance. No one is sure how this works, but the savvy virus takes over the cellular machinery and transforms it into an assembly line that can produce up to ten billion new viral particles every day. Because each replication generally produces a couple of new mutations, the resultant HIV viruses become a swarm of mutational variants. HIV's enormous diversity is the crux of the immune system's challenge.

With enough time, the virus ultimately prevails as the hapless victim's immune system loses its punch. Eventually, the CD4-T-lymphocyte population drops below 200 CD4 cells per cubic millimeter of blood. The CD4-T-cell decline is the hallmark of AIDS-mediated immune depletion. Usually within months of the CD4 cell crash, the patient succumbs to cancers or to one or more devastating microbial infections. Microbes that are easily cleared by healthy people become fatal to AIDS patients.

In the majority but not in all of HIV-infected individuals, just before the collapse of the CD4-T-cell population, the constantly mutating virus undergoes a series of specific mutational changes that alters its cell receptor preference. The HIV *env* gene, which specifies the sugar-coated viral surface glycoprotein, becomes altered so that the virus now binds more effectively to a different chemokine receptor, CXCR4, instead of CCR5. The HIV population within the patient becomes dominated by CXCR4-utilizing HIV variants, ones that enter cells with CD4 and CXCR4 receptors on their surface, but will not use CCR5. The CXCR4 viruses are more virulent than their predecessor CCR5-utilizing viruses. The CXCR4-HIV isolates kill the

T-cells they infect and cause the newly infected cells to produce toxins that kill other yet-to-be-infected T-cells. The CCR5-CXCR4 switch is an ominous sign that previews the rapid depletion of the CD4-T-lymphocytes in the final act of immune destruction.

An important experiment that helped reveal the critical role of CCR5 and CXCR4 involved showing that the specific natural chemokines, which normally bind to the CCR5 and CXCR4 receptors, could physically block HIV infection of susceptible cells. So if a cell with CCR5 and CD4 were first saturated with one or more of the chemokines that normally bind CCR5 (named RANTES, MIPIα, and MIPIβ), the cells would no longer allow CCR5-HIV infection. Their receptor was physically blocked. The same was true for CXCR4-HIV isolates. They could not infect CXCR4/CD4-bearing cells when the CXCR4-specific chemokine ligand—named "stromal derived factor," or SDF—was poured on the cells first. These experiments proved that chemokine receptors were the principal entry portal for HIV infection. But they also made me wonder whether the course of HIV infection could be altered by variants in the human genes that made CCR5, CXCR4, or their counterpart chemokine molecules.

The rapid-fire revelations of the cellular culprits that empowered the AIDS virus galvanized my research team. DNA assays were designed the next day to look for human variants in *CCR5, CXCR4* and all their complementary chemokine genes. (Gene names are usually italicized, while the protein they encode is not; so the *CCR5* gene specifies the CCR5 protein.) We found two common DNA variants within the *CXCR4* gene and a few others in several chemokine genes. But alas, their allele and genotype frequencies were no different among the clinical disease categories; the mutational variants were irrelevant to AIDS.

Then, on Independence Day, July 4, 1996, the day the movie of the same name made its debut, Mike Malasky and Mary Carrington, two talented new recruits to our AIDS project, screened the *CCR5* gene of forty patients. They discovered a few people who carried a dramatic genetic difference: a CCR5 gene with a gaping hole in the middle, a thirty-two-nucleotide-letter deletion. The variant was unusual because it was not a single letter change like most single nucleotide polymorphisms (SNPs), so common in human chromosomes. Instead, the coding script of the CCR5 gene had thirty-two letters missing entirely, snipped out by some long-ago mutation. The variant was found as a heterozygote (i.e., the patient carried one normal CCR5 allele and one mutant allele called CCR5-Δ32) in one-fifth of our patient population. This rather high incidence of the variant was a surprise because the gene product of CCR5-Δ32 is foreshortened and not at all functional. In fact, we now know that the CCR5-Δ32 variant produces a protein so damaged that a cell's garbage-removal detail simply chews it up and destroys it before it can reach the cell surface. People with two copies of CCR5-Δ32/Δ32, about 1% of white Americans, have no CCR5 chemokine receptors on their cell surface at all.

Working around the clock, Carrington and Dean went to our massive patient DNA collection and within a few days they had CCR5 genotypes for 1,955 AIDS cohort participants at risk for HIV infection.

The results were breathtaking. The highly exposed individuals who had avoided HIV infection included the three CCR5 genotypes we expected: people with two normal alleles, CCR5-+/+; those with two CCR5 mutant CCR5-Δ32 alleles; and those with one normal and one mutant allele, CCR5-+/Δ32. The shock came when we inspected data from the infected group; they had only two CCR5 types— CCR5-+/+ and CCR5-Δ32/+. The third genotype, with two copies of the defective gene CCR5-Δ32/Δ32, was never found among 1,343 HIV-infected patients. The implication: People who inherit two copies of the CCR5-Δ32 deletion, one from each parent, are completely resistant to HIV infection. They never get infected even if they are highly exposed over and over again. The mutation removed HIV's only doorway, conferring on its lucky carriers genetic resistance to HIV.

The frequency of the CCR5-Δ32 allele was 11% in our cohorts— predominantly American gay men and hemophiliacs of European descent. The frequency of CCR5-Δ32-bearing people is much lower in African Americans, around 1.7%, and the allele is completely absent in native African peoples, a point I shall return to shortly. The Caucasian incidence of homozygous genotype ($CCR5$-$\Delta32$/$\Delta32$) was between 1% and 2% among uninfected people. Yet it was flat zero in the HIV-infected patients. Since our original study, over twenty thousand individuals have been genotyped in our lab and in several others, and the results have been sustained. Infected patients virtually never have a $CCR5$-$\Delta32$/$\Delta32$ genotype. The reason is straightforward. $CCR5$-$\Delta32$/$\Delta32$ homozygotes produce no CCR5 on their cell surfaces, no required HIV receptors. The doorway, the elevator down into the subway, is shut down.

Our crisp report of $CCR5$-$\Delta32$ mediating HIV resistance appeared in *Science* in September 1996. But we were not the only ones to discover the mutation. William Paxton, working with Richard Koup and Ned Landou in New York, had discovered the same mutation in two gay men who had mysteriously avoided infection in spite of their admission of multiple episodes of risky sex with HIV-infected partners. Paxton was unable to infect their blood cells with HIV in his laboratory, prompting him to discover $CCR5$-$\Delta32$ independently from us. I first learned of their discovery from a news story on my car radio after our paper had been accepted to *Science* in August. They had reasoned that $CCR5$-$\Delta32$ was a genetic shield for the two men; our report, standing on nearly two thousand AIDS patients, proved it beyond any doubt.

At last, our first AIDS restriction gene, one with a dazzling effect: It protected homozygous carriers from HIV infection—completely, absolutely. Well, actually not quite! The $CCR5$-$\Delta32$ discovery encouraged other AIDS researchers to search through their patient collections, and within the next few years a small handful of homozygous $CCR5$-$\Delta32$/$\Delta32$ patients were found to be HIV-infected. However, when the virus in these unfortunate folks was examined, they were infected with the late-stage, CXCR4-utilizing HIV strains. That strain seldom establishes a primary initial infection because nearly everyone's healthy immune systems can effectively dispatch it, unlike the CCR5-utilizing HIV, which sneaks by most people's immune defenses. However, a whopping dose of a CXCR4 virus, as has occurred in these exceptional people, can gain a foothold in very rare circumstances.

In spite of these very rare exceptions, we were gratified to uncover a powerful natural restriction gene that could block HIV infection. The next step was to ask whether the $CCR5$-$\Delta32$ variant had an effect in individuals with one normal $CCR5$ and one mutant $CCR5$-$\Delta32$ allele. These "heterozygous" patients do get infected, since a good fraction of the HIV-infected individuals had the genotype. However, when we examined how fast infected patients with different $CCR5$ genotypes would develop full-blown AIDS, we found that heterozygotes postpone the onset of AIDS for two to four years longer than people with two normal alleles. $CCR5$ heterozygotes have only half the quantity of CCR5 receptors on their cells compared to nonmutant people. That reduction apparently is enough to retard the replication and spread of HIV in infected individuals, in effect slowing down the progression of AIDS. It was not a huge effect, but a three-year delay in AIDS onset could mean a lot to an infected victim.

When Mike Dean looked at the kind of AIDS-defining diseases that patients carrying $CCR5$-$\Delta32$ developed, he found that one AIDS outcome, B-cell lymphoma, a common cancer in AIDS patients, was cut in half among $CCR5$-$+/\Delta32$ heterozygotes. The CCR5 molecule is expressed normally on the surface of B-lymphocytes, a type of white blood cell involved in antibody synthesis, where the lymphomas usually originate. We believe the reduction of lymphoma frequency by $CCR5$-$\Delta32$ may mean that HIV interacts directly with the CCR5 receptor on B-cells as an early step in the induction of these fatal cancers.

Across two decades of AIDS, a small fraction of individuals, no more than 5%, have avoided all signs of AIDS despite their being infected by HIV. How these long-term survivors have dodged the HIV bullet is still a mystery. Are they carrying an HIV strain that has become genetically attenuated? Harvard's Bruce Walker and Janis Georgi of UCLA reckoned that some immune systems are simply better or more effective than others. But why? Is it because of a previous virus exposure that boosted their immune system? Or did they simply inherit a better immune response? Getting at these questions increased our resolve to search for genetic explanations for alternative clinical outcomes to HIV infection.

The horror of AIDS ravaging through Africa, the West, Asia, and the world stimulated researchers to develop larger and more extensive AIDS cohorts. Our own group tested more and more gene candidates that, like *CCR5*, might make a difference to their carriers. *CCR5-Δ32* was intriguing, but certainly it was only part of the answer. Between 80% and 90% of people who inexplicably avoided HIV infection in the face of clear exposure did not carry the protective *CCR5-Δ32/Δ32* genotype. How did they avoid infection? And less than 10% of the long-term survivors, HIV-infected but still healthy for decades, were *CCR5-Δ32* carriers. We surmised there must be other AIDS restriction genes, and we guessed right.

Close to sixty chemokine genes occur in any person's genome, along with a dozen genes for large chemokine receptors that, like CCR5, span lymphocyte cell membranes. Shortly after *CCR5-Δ32* was discovered, Mike Dean found a single nucleotide letter variant in another chemokine receptor, CCR2, a protein that a few rare HIV strains use to enter cells. The variant specified a rather innocuous DNA substitution that changed the coded amino acid from valine to isoleucine, a relatively minor chemical change for the protein. The variant amino acid was physically situated within the membrane, spanning part of the CCR2 protein, and we really did not expect it to have any effect on HIV. We were correct for HIV infection, but we were surprised to see that individuals carrying one or two copies of the variant (called *CCR2-64I* for an isoleucine substitution in the sixty-fourth amino acid position) postponed AIDS symptoms two to four years longer than people with the normal *CCR2* genotype. The *CCR2-64I* effect was equivalent in its AIDS protective influence to the heterozygous *CCR5-+/Δ32* genotype.

The frequency of *CCR2-64I*-protected individuals among our cohort populations was 18%, while the *CCR5-Δ32*-protected carriers composed around 20% of the study participants. So, between 35% and 40% of the HIV-infected patients had one or the other genetic protection, quite a large number we thought. Although the two AIDS restriction genes were rather common, their strength of protection against AIDS progression was slight, only a two-to-three-year delay of the inevitable immune system breakdown. Nonetheless, the AIDS cohort approach was working, revealing small cumulative genetic influences in the patient populations.

Unfortunately, the CCR5-Δ32 variant is completely absent in native African and native East Asian peoples. This is because the original mutation occurred just once, some time after the ancestors of the first Europeans had migrated from Africa to the Northern Hemisphere. The CCR2-64I variant also has an interesting frequency distribution among different ethnic groups. CCR2 genetic protection was present at more than twice the European frequency (10%) in native Africans. In Africa the CCR2-64I allele frequency is 23%, translating to a CCR2-64I heterozygous carrier frequency of 35%. (The heterozygous carrier frequency in a population follows a statistical distribution equal to two times the frequency of both alleles; that is, $2 \times 0.23 - 0.77 = 0.35$, or 35%.)

When we screened CCR2 genotypes in a cohort of African prostitutes in Nairobi who had become HIV-infected, the CCR2-64I carrier incidence was near 50% in the long-term survivors who avoided AIDS, compared to a incidence of 18% in those who rapidly succumbed to AIDS. The CCR2-64I allele seemed to delay the onset of AIDS in these African women for twice as long as it did in the American cohorts. Could it be that the much higher frequency of the CCR2 protective variant allele, plus its much stronger protective effect, reflects the African population's gradual adaptation to HIV in the absence of CCR5-Δ32 protection? We are not sure this is so, but the numbers certainly point to such an explanation.

The mechanism of CCR2-64I-mediated protection was originally a mystery since the vast majority of HIV strains utilize CCR5, with less than 10% using CCR2 as an entry portal. Today the experimental evidence seems to point to an indirect mechanism for CCR2-64I action. It seems that CCR2-64I proteins bind a bit more strongly to CCR5 molecules inside cells and tie them up on their way to cell surfaces. So CCR2-64I variant gene products can function perfectly well in their chemokine receptor function for assuaging bruises, but as a bonus they slow the CCR5 molecules' journey to cell surfaces. The

CCR2-64I-mediated reduction of CCR5 doorways for HIV retards virus spread in *CCR2-64I* carriers, delaying AIDS progression.

Almost as rapidly as we could identify them, new genes that slowed or sped up AIDS began to appear in our cohort gene screens. At this writing, we have confirmed no fewer than twelve different AIDS restriction genes, all common variants that play a role in HIV infection, in AIDS progression, or in the conditions that define AIDS. Two are mutants in genes for specific chemokines: RANTES, which binds CCR5, and SDF1, which binds CXCR4. These two gene variants slow AIDS, apparently by overproducing the chemokines that bind available receptors, physically blocking HIV cell entry and spread in the body. Another AIDS restriction gene variant was found in the gene that codes for a powerful cellular factor named interleukin 10, or IL10. Large amounts of IL10 inhibit the growth of macrophages, monocytes, and HIV. Different nucleotide letter variants in the promoter, rheostat switch region of the *IL10* gene will slow or accelerate AIDS progression by altering the concentration of these cellular molecules.

Three AIDS restriction gene variants involve the human major histocompatibility complex, *HLA*, a dense cluster of 225 different human genes on chromosome number 6, many of which facilitate immune responses to infectious agents like HIV. The *HLA-A*,-*B*, and -*C* genes make a cell surface protein that grasps small peptides made by foreign viruses. Then the HLA-peptide complex alerts the immune system to clean out the viruses. These *HLA* genes are well known as possessing enormous allelic diversity in populations; over four hundred human alleles exist at just three loci: *HLA-A*, -*B*, and -*C*. The extensive variation is part of an evolutionary strategy to provide a highly diverse repertoire for recognition of foreign agents. People and species with low MHC variation succumb to viruses more readily. The cheetahs' hypersusceptibility to the feline peritonitis virus, described in Chapter 2, provides a cogent example of MHC-mediated vulnerability to fatal viruses.

It also turns out that like the cheetah, people with limited MHC variation are at a severe disadvantage in the face of HIV. Homozygotes for *HLA-A*, -*B*, and -*C* (i.e., individuals with two copies of the same allele at one or more of the *HLA* genes) make up a large part of the rapid progressors to AIDS, the unfortunate victims who develop AIDS within two or three years of infection. The explanation is that rapidly mutating HIV simply evolves its own resistance to the *HLA* defenses faster in patients with less *HLA* variation.

Our grueling twenty-year monitoring of thousands of AIDS patients was beginning to bear fruit. The genes that orchestrate the victim's unwitting collaboration in the march to AIDS had reared their heads after eluding discovery for so long.

As the involvement of AIDS restriction genes came into focus, we had a practical concern. How important were these genetic variants anyway? In the big picture of a global plague, did they really matter? Some of the effects were pretty small, revealed only by the scale and diagnostic precision of the AIDS cohort population. Was there a way to grasp their real influence? And how much of the differences in AIDS epidemiology could a person's genotype influence?

Answering these questions can be complicated because the physiological actions of the genes are interactive, so we cannot simply add the effects of each gene together. Epidemiologists can, however, quantify the influence of a genetic risk factor by considering three aspects of AIDS restriction. The first is whether the gene can exert its effect in one dose (i.e., it is dominant) or requires two doses (it is recessive). For example, *CCR5-Δ32* exerts a dominant (one dose) restriction on AIDS progression and recessive (two dose) effect on HIV infection. Second, we consider how strong the restriction gene's effect is. How much better off are people with protective genotypes than those who lack them? This aspect is called "relative risk" and is measured by the ratio of protected versus susceptible genotypes among people who develop (or resist) AIDS. The third consideration is the frequency of the protective genotype in the population at large. Common protective genotypes are more important to an epidemic than rare ones.

Epidemiological theory allowed us to assess the influence of each of the twelve known restriction genes separately and then as a group, since the very same patients were examined for every gene. Each of the twelve genes has some measured effect on the rate of AIDS

progression. The estimated influence of any one-restriction gene separately was generally small; for example, somewhere between 5% and
10% of the long-term survivors can attribute their delayed onset to
CCR5-Δ32. However, when all the AIDS restriction genotypes are
considered as a single factor, about a third of the epidemiological
variation in the rate of AIDS progression was attributed to a victim's
genotype. Put another way, about 15% of the long-term survivors and
15% of the very rapid AIDS progressors are in those disease categories as a consequence of the form of AIDS restriction genes they
carry. This means that 15% of the very fortunate long-term AIDS
survivors, the people who avoid AIDS for twenty years, would have
succumbed much sooner were it not for genetic protection by one or
more of the dozen AIDS-restricting genes discovered so far. The
same is true for rapid progressors: 15% would have survived much
longer if they had received a better genetic deal.

The high extent of cumulative genetic attribution spotlights the
critical role genes can play in controlling the speed with which different people succumb to AIDS. We have less knowledge of the genetic
influences for the initial HIV infection process, since only three of
the eleven genes, *CCR5, RANTES,* and *IL10,* show an effect on HIV
transmission. I expect there are many other undiscovered restriction
genes, some weak, others potent, that will be shown to regulate or
limit HIV infection, or progression to AIDS. Today we are searching
the AIDS cohorts for new restriction genes that regulate immune
response to HIV and sensitivity to the powerful anti-HIV drugs.

The AIDS restriction genes do not provide an immediate cure, a
vaccine, or even certain diagnostic prognosis, but they put our genes
in the mix and start AIDS researchers toward a different line of
thinking.

It is perilous to predict confidently a tangible medical benefit for
new discoveries. Some things simply do not work out. Nonetheless, if
one could forecast a payoff for our investment in AIDS restriction
genes, it would be a translation of the gene's mechanism into innovative and effective therapies that would mimic natural stalls on HIV
and AIDS. Almost immediately after CCR5's role in HIV infection

was revealed, pharmaceutical companies began exploring ways to exploit HIV's critical need to enter cells through CCR5. If most people's healthy immune system can dispatch the CXCR4-utilizing HIV, then a drug that halted the more successful CCR5-mediated HIV cell entry might be an effective block to HIV infection.

Small synthetic peptides that bind to CCR5 but do not stimulate intracellular gene signaling (the normal function of chemokine receptors) are showing extraordinary promise in animal studies and in human clinical trials. The therapeutic peptides physically block the virus's access to CCR5, preventing uptake of HIV by susceptible cells and interrupting AIDS progression. Other compounds such as monoclonal antibodies, which sheath CCR5, or compounds acting within cells to impede CCR5's transport to the cell surface (like *CCR2-64I*) are being evaluated for possible anti-AIDS therapy. The hope is to foul up, with "smart drugs," the cellular machinery that HIV requires to cause damage.

Attacking the host facilitation of HIV makes sense for another reason. The fifteen anti-AIDS drugs already licensed attack HIV genes, interfering with virus assembly. HIV's high mutation rate—nearly a billion new mutations each day—will allow it to evolve resistance. HIV is a moving target, while cellular genes are not. A treatment that attacks the cellular gateway might not be compromised by rapidly evolving viral resistance.

It is unimaginably lucky that people homozygous for *CCR5-Δ32/Δ32* are perfectly healthy. CCR5 is dispensable because its important chemokine receptor function is genomically "redundant." From a therapy development perspective, it is remarkably good fortune to uncover a gene that is absolutely necessary for a fatal infectious disease to progress, but otherwise expendable. Such a function is an ideal drug target.

CCR5 also happens to be part of a large family of proteins, called "seven transmembrane spanning receptors," that are very familiar to pharmaceutical companies. Drugs that home in on seven transmembrane receptors have already been developed to treat common inflammatory diseases such as asthma, ulcers, arthritis, and psoriasis.

Hundreds if not thousands of anti-seven-transmembrane-receptor drugs have already been tested for human toxicity and effectiveness. Now these are being dusted off and tried again for HIV.

At least two promising drugs that block HIV-CCR5 interaction are now entering the final stages of human clinical trials, the last step before they can be licensed by the FDA for public use. The incremental task of translating basic research to the bedside is playing itself out. The proof of these bold new therapies must await their application to the epidemic. Until a cure and an effective vaccine emerge, the slow, patient research, the trial and error, will continue.

The face of the AIDS epidemic has many expressions. I personally witnessed the horror of the disease through the eyes of my only brother as every afflicted life in his world was shifted to "fast-forward." The mind-numbing 80% incidence of HIV infections in the 1980s gay community of San Francisco was marching countless young men in their prime to an early grave. Gradually, I became accepted in their company as a conduit to AIDS research with the NIH medical community. I explained the newest discovery, the latest advance, any glimmer of hope. All the while, I knew that it was only a matter of time until I would bury yet another AIDS casualty among my new friends and their partners. Seeing the steady progression of AIDS so closely emboldened my resolve to hasten our research toward an answer, to turn the deadly scourge. Today, we continue our search for undiscovered AIDS restriction genes by mounting full human genome "scans" involving tens of thousands of gene variants, hoping to reveal all the AIDS-limiting genes there are.

The AIDS story is far from over. Most observers familiar with the pace of AIDS research are cautious in predicting how this episode will play out. No scientist or AIDS advocate that I know is resting easy or counting on a quick solution. In fact, many believe that the nonscientific public will not forgive us if we let up in our vigilance. AIDS has killed nearly as many people in two decades as did the Black Death, more than the flu epidemic of 1921, and more than smallpox introduced to Native Americans by Cortés's army.

There is little time for celebration of modest advances as the epidemic soars. One can only hope, work, and pray that a vaccine, a cure, an intervention is developed before the infectious disease becomes so rampant that virtually all of the human species are exposed. Should that unthinkable prospect bear out, then scientists will no longer view AIDS as an infectious disease but as a genetic disease suffered by those unable to combat inevitable exposure to a ubiquitous HIV. Then the virus will have won.

Thirteen

Origins

SHE WAS TREMBLING, FEVERISH, AND AFRAID. IT WAS JUST before dawn and she had suffered a fitful night, dozing, praying, but mostly crying. She prayed so hard that she developed a throbbing headache and was slipping toward hopelessness. Over the past weeks Margaret Blackwell had helplessly watched her son, her two daughters, her uncle, and her cousins wither and perish from the dreaded pestilence. She lived in the tiny English hamlet of Eyam in Derbyshire. Margaret was thirty-six; the year was 1666.

In the face of the growing plague, the twenty-eight-year-old vicar of Eyam parish, William Mompesson, a devout and domineering prelate of the Church of England, pronounced a cordon sanitaire, a strict quarantine for the eighty parish families to their homes in Eyam to limit the epidemic's spread. Food provided by the Earl of Devonshire was delivered to a remote junction at the southern limits of the village, exchanged for the villagers' meager coinage soaked in vinegar. Mompesson announced that the church would be closed and services held outdoors. Burials and funerals were stopped, forcing villagers to bury their loved ones in their fields and gardens. The terrified citizens agreed to isolate themselves for the good of others as prescribed by their pastor.

The minister could not have known, but his decrees were worse than useless. Fleas lurking on the fur of village cats and black rats readily spread the bubonic plague. The vermin would transport the disease from house to house as freely as the sun would shine through the windows. In a twinkle of historic time, every devout citizen of Eyam who remained (a few had fled in horror) became exposed to *Yersinia pestis,* the plague bacillus, the same deadly bacterium that had ravaged Europe three centuries earlier, causing the Black Death.

Margaret Blackwell pulled herself from her bunk and hobbled down the loft ladder to warm her trembling body near the slow-burning hearth. She tried to move quietly so as not to disturb her mother and brother, who were mercifully still sleeping. Margaret dared not question the wisdom of the Creator who was taking so many lives in such a horribly painful manner. She had heard of the Black Death ages earlier and of supplicants who would flagellate themselves in public piazzas of France and Italy to appease God's ire. She had prayed so earnestly her knees were bruised and inflamed. Or was the tissue inflammation part of her disease? She spotted the urn of bacon fat on the mantel saved for cooking; it was still tepid and fluid. In her delirium she thought the cooking fat would surely kill the demons that had possessed her frail and dying body. Maybe it would finish the evil within her, or at least shorten the agony. She drank the whole jug.

Before she could even contemplate her recklessness, she doubled over and began to vomit, to screech, to wail, and to whimper. Exhausted, she passed out, certain it was her last breath.

But it was not. In a few days her strength slowly returned, her fever subsided, and she rose to nurse her stricken family and neighbors.

The death toll in the tiny village was staggering. Of eighty households, seventy had at least one victim; most had more. The burials would reach 280 by the end of 1666, over 50% of the populace of Eyam. My visit last year to the Eyam churchyard provided a grim image of those horrid years, rows and rows of headstones bearing the years of death 1665 and 1666—the plague stones.

A hundred miles to the south in London, tens of thousands succumbed to the massive seventeenth-century plague. In the spring of 1666, the great fire of London probably played a role in ending the dying. Yet, not everyone who contracted the disease had perished. What had saved them? Was Margaret Blackwell cured by the whopping dose of bacon fat she ingested? Blackwell's great-great-great-grandniece, Joan Plant, is today administrative director of the Eyam parish church. Joan, who told me this story, thinks the radical potion might have spared Margaret. When we met, I had another idea.

It is difficult to imagine the horror that fourteenth-century Europeans endured as the Black Death marched across Europe. Historians believe it began in the early 1300s somewhere in Southeast Asia— perhaps on the Mongolian steppes, in the Himalayan valleys, or even in Burmese ghettos. A particularly virulent strain of the *Yersinia* bacterium emerged from a less lethal form endemic to rodents. The microbe thrives in the blood of a dozen rodent species and had adapted itself to transport by fleas.

The earliest Asian mortalities were recorded in the Gobi Desert of central Asia, where marmots (large rodents related to woodchucks) carried the plague organisms. Trappers collected the dying marmot pelts, laced with hungry fleas, and sold them to dealers for transport west over the Silk Road to Kaffa, a bustling seaport on the northern coast of the Black Sea. The dense human population of Kaffa, abominable hygiene, and a flourishing black rat presence provided an ideal condition for transmitting the hearty plague bacillus. The rats freely boarded medieval ships and reached the teeming Mediterranean port city of Messina, Sicily, in October 1347. When the sailors began to off-load the corpses, they and their ship were quickly dispatched from the harbor, but it was too late to stop the rats from running ashore. The course of Western civilization would be changed forever.

The plague-afflicted victims first developed large egg-sized swellings in the lymph nodes of the groin and armpit. These are the buboes from which the bubonic plague derives its name. Within a few days of the appearance of buboes, the pitiful sufferers acquired high fever, delirium, and hemorrhagic black splotches, the mark of the plague, or God's "tokens," all over the body. Internal bleeding then led to neurological toxicity; internal organ breakdown; bleeding from the skin, bowels, and nasal passages; and finally, a horrible, painful death.

The infecting microbe, *Yersinia pestis,* was first implicated as the responsible agent for Black Death in 1894 by Alexandre Yersin and Shibasaburo Kitasato. The bacterium has evolved a formidable arsenal of killing machinery in the form of an extra-bacterial chromosome (called a "plasmid") that encodes a repertoire of toxic proteins called "*Yops*" (Yersinia outer proteins). Yops poison cellular machinery, particularly that of circulating lymphocytes that mediate immune defenses. Once in the bloodstream of their host, the bacteria home in on macrophages, an early bacterial defense cell, puncture holes in its surface, and inject six *Yop* proteins that destroy the macrophage before it can send a chemical distress signal to the other arms of the immune system. The victim's defenses neatly destroyed, the host becomes a dying but efficient factory for making *Yersinia* bacteria. Today, we can treat the one thousand or so cases of plague that occur worldwide each year with powerful antibiotics, but such compounds did not exist in the fourteenth century.

The Black Death devastated European communities, spreading from city to city with unspeakable mortality. When it hit, the population of Europe was around 100 million. Within five years, 30–40 million people had perished. Half the population of Italy and England succumbed; in Venice, three-quarters of the population, 100,000 people, died. Eighty percent of the Genoese populace was lost. From London, reports of mortality as high as 90% were recorded. The islands of Cyprus and Iceland were said to have been completely depopulated by plague.

By 1352, the plague had marched north from Italy through France, Germany, England, and Spain, on to Scandinavia, and then west to Russia to return to within a few hundred miles of where it began near Kaffa. Thirty million Europeans had been lost. Then the dying stopped as abruptly as it began.

Less precise are the estimates of the Asian mortalities that preceded the European plague. However, many historians assess an even greater loss of life in Asia than in Europe. Daniel Defoe, best known for *Robinson Crusoe,* wrote *A Journal of the Plague Years* in 1722 in which he described fourteenth-century India, China, and Asia Minor as literally covered with dead bodies. The culturally advanced Chinese population would be cut in half by the plague and subsequent famine, from 123 million in 1200 to around 60 million in 1350.

The Black Death was not the last epidemic of plague to cripple Europeans. The horrific disease reappeared ten years after the 1348 wave and with almost equal intensity. Periodic regional outbreaks would occur with alarming frequency and intensity once every generation for the next three hundred years. In the centuries following the Black Death, European population declined 60–75%. The Great Plague that swept the British Isles in 1665 claimed seventy thousand lives, including the terrified citizens of Eyam parish. The last major plague outbreak occurred in Marseilles in 1772, when half of the city's population fell to the dreaded disease.

The history sections of university libraries abound with books about the Black Death and its social consequences. Suffice it to say, everything changed in its wake. People became grim and more introspective, emphasizing individuals over communities. They embraced a certain fatalism in contrast to the hopefulness of devotion. The very pillars of religious institutions were challenged. How could a merciful Creator inflict this sort of damage? The most common conclusion was that God's wrath was punishing His people for their sins, and severely.

But other explanations emerged as well. Most realized that the disease was transmitted from person to person, but no one was sure how. Some Christians put the blame on the Jews, who by then had already been subjugated to second-class citizenry or even servitude by the Church. Pope Innocent III labeled them "Christ killers," and Thomas Aquinas reasoned that "since Jews are the slaves of the church, she can dispose of their possessions." Rumors flew that Jews did not seem to be as susceptible to the plague, and some reasoned that they had hatched an insidious plot to poison the water wells, causing the blight. The rumors led to terror.

In Mainz, Europe's largest Jewish stronghold, a mob of vigilantes overpowered the terrified Hebrews and burned six thousand to death on August 24, 1349. At Wormes in March of the same year, four hundred Jewish citizens set themselves afire in their own homes to avoid being killed by Christians. Three thousand more were killed at Erfurt shortly thereafter. All told, an estimated sixteen thousand Jews were brutally murdered in those years as a racist solution to the Black Death.

The Black Death of the fourteenth century was not the first wave of plague to curse Europeans. During the sixth-century reign of the Byzantine emperor Justinian, the first major emergence of plague is recorded. The ferocity of that epidemic was horrific, at its height killing ten thousand people a day in Constantinople. The Justinian plague spread northward through Europe and the British Isles, decimating the Roman Empire. Afterward, waves of plague cycled through Europe with alarming regularity until the late eighth century, when the dying finally ceased. Modern estimates place the death toll of the Justinian-era plague (541–750 A.D.) on the order of 100 million people, a number that surely overwhelmed the Roman numerals in use at that time.

The Justinian Plague, the Black Death, and the pestilence that followed went unabated by treatment, by quarantine, or by the medical wizardry of the time. The exposed became infected, nearly all got sick, and 60–80% died. The horror was chronicled by clerics, theologians, novelists, artists, and historians; but their view was limited to descriptions of the suffering. Medicine was theology; microbiology did not exist and molecular genetics was not even imagined.

Many questions about the plague remain. Why did it start? Why did it end? Why did some die and others survive? There are serious disagreements as to the cause. Monographs have been written that attribute the Black Death to anthrax, measles, typhus, and even tuberculosis. *Biology of Plagues*, a 2001 book by British researchers Susan Scott and Christopher Duncan, argues that the Black Death was caused by a hemorrhagic fever virus similar to the deadly Ebola or Marburg viruses. Most experts are still confident that *Yersinia* is the responsible agent. Indeed, DNA technology recently uncovered *Yersinia* DNA in the dental pulp of fourteenth-century plague victims buried in French grave sites. Nonetheless, scientific queries and uncertainties are thought-provoking and important, particularly if we hope to avert a similar calamity in the future.

Now fast-forward to the most formidable of modern plagues, the scourge we call AIDS. For that menace we actually do have living patients, medical surveillance, tissue specimens, and sensitive molecular tools that led scientists to the viral cause, HIV. A billion-plus-dollar annual AIDS research expenditure has allowed scientists to dissect HIV's mode of action and to trace the spread of HIV across the planet. Unfortunately, as advanced as our biotechnology has become, we have yet to design a preventive vaccine or curative treatment.

With 60-plus million people infected so far and more than 22 million deaths in the two decades since it was first recognized, HIV's mortality numbers are getting uncomfortably close to those of the Black Death. A continued rate of 6 million new infections per year would project AIDS to eclipse the 30-million death toll of the fourteenth-century plague in the next few years.

In spite of these depressing statistics, science has learned a lot about the origins of HIV and AIDS. If we compare HIV genetic changes across time, across the continents, and across animal species that carry related lentiviruses, a clear view of how and when AIDS first entered humankind has come into focus. That story, affirmed and crystallized only very recently by combining medicine and genetic data, goes something like this.

The AIDS virus HIV comes in two genetically distinct types, HIV-1 and HIV-2. These two strains are distantly related to each other and it turns out they have different origins. HIV-1 is the potent and ubiquitous virus that is responsible for the global AIDS epidemic. HIV-2 is a less aggressive strain found largely in people living in West and central Africa. HIV-2 will cause an immunodeficiency and wasting disease, but far more slowly and less efficiently than HIV-1.

Blood screens of nonhuman primate species in Africa have revealed over twenty different wild monkey species that harbor simian immunodeficiency viruses (SIV), genetically the first cousins to HIV. These SIV-infected wild monkeys are dispersed throughout

Africa, and they tolerate the SIV infections without developing AIDS, an interesting situation to which I will return shortly.

Careful phylogenetic analysis of the simian virus genomes showed that HIV-2 has its origins from the endemic SIV circulating in sooty mangabeys (*Cercocebus atys*) in Africa. The more widespread and highly virulent HIV-1 strain that has traversed the globe descended from a virus circulating today in wild chimpanzees (*Pan troglodytes*) in West and Central Africa. The earliest HIV-infected human blood specimen was taken in 1959 from a British sailor who spent time in the same region of Africa.

Epidemiological studies show that the first instances of AIDS-like diseases appeared in the Congo and in neighboring countries of central Africa, close to the region where the chimpanzees live. Comparisons of several hundred HIV and SIV genome sequences have revealed a pattern of viral relatedness that is best explained by no fewer than seven independent transfers of monkey SIV into human victims. Three distinct HIV-1 strains jumped into humans from chimpanzees, but at different times. Two of the virus transfers originated from the West African chimpanzee subspecies *P. t. troglodytes* and a third from the central African chimpanzee subspecies (overlapping western Congo, Tanzania, and Uganda) *P. t. schweinfurthii*. The first virus then spread beyond Africa, while the second two strains remained and percolated in central African communities.

Four separate transfers of SIV from sooty mangabeys to humans also happened—at least that is how we interpret the phylogenetic cluster patterns of the viral genome sequences. The sooty mangabey virus, termed HIV-2 once it got into humans, spread among several West African cultures, primarily through heterosexual sex, particularly in people with multiple partners.

As much as can be determined, it appears that neither chimpanzees, sooty mangabeys, nor the other monkey species infected with their particular SIV strains develop AIDS-like illnesses themselves, even when they carry the virus for years. Apparently these SIV-infected monkey species have already passed through a historic adaptive episode.

Scientists feel certain that the wild African monkeys are genetically resistant, and not simply infected with a virus that lost its punch, for two reasons. When SIV from prospering African monkeys is transferred to Asian macaque species, these monkeys develop AIDS quickly. Asia is free of SIV, so wild macaques had never been exposed. The lesson came to us by accident: When macaques were housed with healthy SIV-infected African monkeys in an NIH-run primate research center, they came down with AIDS, having caught the virus from the healthy African monkeys. Had they not taken ill and died of an AIDS-like immunodeficiency disease, we might still be wondering where HIV came from.

That the chimpanzee SIV is still virulent is also attested to by the ferocity of the AIDS epidemic in humans. HIV-1, the offspring of chimpanzee SIV, exerts over 90% mortality, making it more deadly than any infectious disease ever recorded.

How long ago did SIV enter the human population? Put another way, how old is the human version of this fatal disease? The age of HIV-1 is proportionally related to the amount of genetic diversity seen today among the HIV isolates spread across the world. The virus sequence diversity accumulated over time can be used to estimate when the virus first entered humankind. The most robust calculations indicate that it would take about seventy years to accumulate the worldwide level of variation present in HIV-1 strains today. That means that the first virulent forms leapt from chimps to African victims sometime in the 1930s. We do not have any blood samples with antibodies to HIV from before the late 1950s to prove this estimate. However, the imputed 1930s date seems to fit what data we do have rather well.

But how did HIV-1 actually get transferred from chimpanzee to human? Monkeys, small and large, were and are hunted in a widespread, often illegal bushmeat trade to provision the logging industry as well as upscale restaurants. African bush hunters shoot and butcher hundreds of monkeys each month in areas where populations of chimpanzees, gorillas, and other smaller SIV-infected monkeys reside.

African bush hunters are neither taxonomists nor conservationists, so they seldom discriminate between common small monkey species and rare, endangered species of apes, our closest relatives. Indeed, bush hunting is often tolerated in Congo, Cameroon, and Gabon by government officials in spite of international laws protecting endangered species. We are having members of our own family for supper. A graphic account of the extensive bushmeat trade and its impact on remaining chimpanzee and gorilla populations is chronicled by Karl Ammann, wildlife photographer turned conservationist, and author Dale Peterson in their haunting book *Eating Apes,* due out in 2003.

The chimpanzee's SIV almost certainly made its way into bush hunters through the bloody butchering process. The virus would be transmitted relatively easily through open cuts in the hands, periodontal lesions in the mouth of the diner, or any other blood contact. The initial infections probably caused severe disease in the early recipients, but not before they transmitted the slow-burn virus to their sexual partners. The disease was transmitted at low frequency in central African villages for decades until social conditions there and outside of Africa led to its recent global dispersal.

The first recognized clinical AIDS cases in Africa occurred in the Bukoda and Rakai districts of Uganda and Tanzania, near the western shore of Lake Victoria, in the early 1980s. Prior to that, social conditions catalyzed by the civil war to unseat Ugandan dictator Idi Amin may have invigorated the spread of the deadly disease. Major truck routes passing through the battlegrounds, waves of war-related rape victims, plus the flood of prostitutes fleeing from the war zone to urban brothels facilitated the rapid-fire dispersal of HIV.

Jumbo jets transferred infected people from Africa to Europe, America, and Asia before any broad understanding of the disease existed. With the exception of a few very recent successful health education programs—notably in Uganda and Thailand—the virus continues its accelerating course unchecked in the developing world today.

None of this was supposed to happen. In 1967, U.S. Surgeon General William Steward pronounced an end to the infectious disease era. Encouraged by the successes of antibiotics and vaccines against smallpox, measles, polio, and others, he predicted a swift shift in biomedical emphasis from infectious to chronic diseases. The devastation of AIDS, of hepatitis B—400 million infected and at risk for liver cancer—and of papillomavirus, the principal cause of cervical cancers, has disabused public health professionals of the notion that modern medicine has conquered infection. Not even close.

The abrupt catastrophe of AIDS dwarfed the death toll of other recent disease outbreaks; it should have invigorated international vigilance against the deadly microbes that surround us. In the mid-1990s, near hysteria erupted in Europe over bovine spongiform encephalopathy (BSE), also called "mad cow disease." BSE is a debilitating neurological syndrome caused by an infectious prion agent that infiltrated English cattle, and afflicted a few, perhaps one hundred people, with a neurodegenerative pathology. The year 2001 saw a serious European outbreak of foot and mouth disease, an endemic viral infection of African buffalo and wildebeest that does not actually kill people, but that can diminish yields of domestic cattle stock. These agents have taken only a handful of human lives, but their prospects have altered appreciably the dietary preference and beef consumption of millions of Europeans and even a few Americans.

Now consider the bushmeat industry in central and West Africa. Scores of chimpanzees and gorillas are slaughtered daily in Gabon, Cameroon, and Central African Republic in an eerie scenario that plays out the bloody contact exposure that allowed SIV virus transmission. The great ape killings are illegal; they violate several international treaties to preserve endangered species. Yet, government agencies look away and implicitly endorse noncompliance. Aside from the conservation issues, the practice has already led to transmission of a deadly virus, HIV, at least seven times. These transfers have taken 22 million lives so far, and 15,000 new HIV infections occur daily. By 2005, the UNAIDS program estimates that it will cost us $9.2 billion annually to effectively combat AIDS spread in the world.

Karl Ammann explained to me that bushmeat is not infrequently served at state dinners as well as in restaurants in West African capitals. To his knowledge no poacher has ever been prosecuted for killing chimps or gorillas. The great ape bushmeat has even been transported on occasion by diplomatic pouch to London, to Washington, and to other foreign capitals for state dinners in the embassies. While our border patrols are hypervigilant, even maniacal, over mad cow disease and foot and mouth disease, aloofness over great ape bushmeat trade abounds. Our carelessness has potentially deadly consequences.

Scientists trace the origins of abrupt new epidemics so we might prevent similar catastrophes in the future. We know how AIDS got here and there are laws in place to prevent a repeat performance. But the resolve to enforce those laws is neither apparent nor pressing in the African nations in which the new transmissions are opportune and likely ongoing. This is no longer a scientific curiosity, nor a local conservation misdemeanor. It is a deadly practice that threatens all of us.

Too many pestilences have drastically afflicted humankind since the dawn of civilization, from the Justinian Plague to twenty-first-century AIDS. All these scourges have one aspect in common: different responses from different people. Margaret Blackwell survived the plague. AIDS activist Steve Crohn and thousands of others remain free of HIV because they carry two copies of $CCR5$-$\Delta 32$, one of eleven guardian genes that protect their carriers from AIDS.

Early on I learned that once an answer to a scientific puzzle is resolved, new and often more intriguing questions pop up to further perplex us. Our discovery of AIDS restriction genes, particularly $CCR5$-$\Delta 32$, made me wonder how the human mutational variant got here in the first place, or why it persisted in such a high frequency. Why would a variant that blew out a seemingly important immune function—a receptor for chemokines—be so common among Europeans and European Americans? We understood the origins of HIV pretty well, but what about the origin of $CCR5$-$\Delta 32$?

Several aspects of the $CCR5$-$\Delta 32$ mutational aberration were unusual, and by putting together different pieces of the puzzle, we were able to unveil a remarkable series of formative events. The $CCR5$-$\Delta 32$ gene variant encodes a version of the $CCR5$ chemokine receptor that is missing thirty-two nucleotide letters. The damaged

CCR5 protein is seen as nonsensical by the cell's security detail, which sends a signal to an enzyme clearance brigade to remove it. As a consequence, people with two copies of the *CCR5-Δ32* allele do not have any CCR5 chemokine receptors on their cell surfaces.

The rather unusual nature of the deletion, thirty-two nucleotides in precisely the same DNA position in every person that carries it, plus a complicated analysis of the neighboring DNA sequence variation close to *CCR5-Δ32* on the chromosome, showed that the mutation occurred only once and was then transmitted to subsequent offspring over many, many generations. This single mutational event will become more relevant in a moment.

Because HIV-1 normally enters cells through CCR5, homozygous *CCR5-Δ32* people avoid AIDS entirely because their HIV receptor portal is missing. Yet, people with this genotype show no genetic or immune disability caused by their loss of CCR5 function. The reason for their good fortune is that the job of CCR5 in trafficking lymphocytes to inflamed tissues is backed up by twenty other chemokine receptor genes also found in human chromosomes. For some unknown reason, one certainly worthy of investigation, the *CCR5* gene function is useful, but dispensable. The same is true for the homologous *CCR5* gene in mice. Mice with their *CCR5* gene experimentally "knocked out" are reasonably unfazed by the genetic engineering, and they live rather healthy lives.

The frequency distribution of the *CCR5-Δ32* variant in the world's population now bears some mention. The variant is common among Caucasian Europeans and their descendant Caucasian Americans, varying from 5% to 15% in allele frequency. However, *CCR5-Δ32* is completely absent from native African and native East Asian ethnic groups. African-Americans have an incidence of 2–5%; their few copies of the variant *CCR5-Δ32* derive exclusively from Caucasian gene flow to the African slaves and their descendants since their transport to the Americas.

Anthropologists and molecular geneticists studying human origins have determined that the earliest ancestors of today's major racial groups migrated out of Africa to populate the globe and to displace other early hominids (like the Neanderthals) living there about 150,000–200,000 years ago. Europeans and Asians became separated from each other by continental barriers and developed into modern ethnic groups some 50,000 years later.

Since $CCR5$-$\Delta32$ is found only in Europeans, the unusual mutation must have occurred some time after they split away from their African and Asian forebears, probably within the last 100,000 years. Since that time, human populations in Eurasia have been large, seldom below 20,000 and usually in the millions. If $CCR5$-$\Delta32$ first occurred in a newborn infant in a population of, say, 100,000 people, the $CCR5$-$\Delta32$ frequency the day she was born was 1 in 200,000. Everybody, including the baby with the new $CCR5$-$\Delta32$ mutation, had two copies of the $CCR5$ gene—so there were 200,000 $CCR5$ genes, one $CCR5$-$\Delta32$ and 199,999 normal CCR5 alleles. That baby had one normal version and one disrupted $CCR5$-$\Delta32$ copy of the $CCR5$ gene sequence.

We know from population genetic theory that 99.9% of new mutational variants, particularly those that destroy a good gene's function, disappear from the population in fewer than fifty generations. But $CCR5$-$\Delta32$ did not disappear; it grew. Today, one European in ten carries at least one copy of $CCR5$-$\Delta32$. So how did the $CCR5$-$\Delta32$ frequency rise from 1 in 200,000 to 1 in 5?

The single explanation that fits the data well is that $CCR5$-$\Delta32$ was itself adaptive for people. The variant conferred some sort of advantage—reproductive, survival, or otherwise—to its carriers. An undiscovered environmental condition—evolutionary biologists would call it a "selective pressure"—made the $CCR5$-$\Delta32$-carrying baby and its future offspring more fit, more likely to survive and procreate in the time, place, and environment in which they lived.

Two additional characteristics about $CCR5$-$\Delta32$ that I have not mentioned also point to a selective advantage for the variant. First, the frequency distribution of $CCR5$-$\Delta32$ across Europe occurs in an unusual continuous gene frequency gradient from northern to southern Europe. The highest $CCR5$-$\Delta32$ frequency is found in Scandinavia, Finland, and northern Russia, where the $CCR5$-$\Delta32$ allele frequency reaches as high as 16%. France, England, and Germany hover around 10%; Italy, Turkey, and Bulgaria, 5%; Saudi Arabia and sub-Saharan Africa, 0%. A gradual gene frequency gradient like this is a hallmark for an intense recent selective pressure in the locale of highest allele frequency, the north, followed by spread of the selected allele over time to the south after the mysterious selective event has subsided.

The second phenomenon that implicates the $CCR5$ gene itself as the object of powerful natural selection involves the pattern of mutational variants other than $CCR5$-$\Delta32$ also found in the $CCR5$ coding region. Mike Dean and Mary Carrington scanned several thousand people from all ethnic groups for $CCR5$ mutations and discovered twenty additional, but rather rare, variants. Nearly all of the $CCR5$ variants changed the amino acid or caused a deletion, like $CCR5$-$\Delta32$. Only a few were innocuous, or "silent," with respect to sequence alteration. (A DNA variant that changes a three-letter codon "word" but still encodes the same amino acid is called a "silent" mutation, because it has little if any consequence to the carrier. Silent codon word variants are like synonyms in English, words with different spellings but the same meaning—for example, blood "spattered" and blood "splattered.")

When most other human genes are similarly examined, the pattern of mutational variants is very different. For nearly all genes, 70–80% of new variants are silent. Silent variants are far less likely to be damaging to a gene's function; they are synonyms and make no difference. By contrast, new amino-acid-altering mutations are usually bad or maladaptive. Codon-changing variants seldom persist for very long before they get eliminated by the process of natural selection. The single exception occurs in genes that defend against invading microbes, those involved in immune recognition like the major histocompatibility complex. These genes favor amino acid alterations because multiple variants increase their ability to recognize and dispatch disease microbes. $CCR5$ fit this model. The excess of amino acid altering versus silent variants within the $CCR5$ gene was a signal that natural selection mediated by historic infectious disease outbreaks had strongly favored CCR5-receptor-altering variants like *CCR5-Δ32*.

CCR5-Δ32 is a relatively young mutational variant that somehow got dramatically elevated from less than 1 in 200,000 to today's high incidence, 10% in Europeans. A mysterious but breathtaking fatal infectious disease outbreak that, like AIDS, exerted a huge mortality, and to which *CCR5-Δ32* carriers were resistant, is the only reasonable explanation. Yet, it could not have been HIV, which appeared only a generation ago, too recently to cause the meteoric rise in *CCR5-Δ32* allele frequency in Europeans.

A critical advance that drew us closer to learning the cause of the *CCR5-Δ32* hike came from estimating the age of the mutational variant. To get at that, we needed to look at other variants in the chromosomal region in which the *CCR5* gene lies. *CCR5* is located on the short arm of human chromosome 3 surrounded by hundreds of DNA variants on either side. Any new mutation would be "born" on a single chromosome inherited from the sperm or egg of the first individual's parents. When that "mutant-zero" baby carrier grows up and marries, his or her children inherit the gene variant. Repeat this transmission over many generations and the variant is propagated to future generations.

However, with each generation, *CCR5-Δ32* and its adjacent DNA variants get reshuffled by paired chromosome exchanges that occur frequently during the formation of sperm and egg cells. Over time new variants like *CCR5-Δ32* become randomized with respect to adjacent DNA variants in the population. However, the shuffling process takes many, many generations for variants close to the new gene variant, *CCR5-Δ32*. Geneticists can measure how big the nonrandom segment of adjacent variants around *CCR5-Δ32* is and then use that length to estimate how much time has elapsed since the allele variant was born, or, more appropriate in this case, when it was most recently elevated by natural selection, that is, the responsible epidemic. The date for the *CCR5-Δ32*-bearing segment computed to 680 years ago—smack in the middle of the fourteenth century.

The *CCR5-Δ32* mutation grew up in Europe during the Black

Death. Could it be that the same mutation that protects against HIV may have also protected medieval Europeans exposed to plague? The timing is right, the numbers are right: Hundreds of years of unforgiving selective pressure from the sixth to the eighteenth centuries would be sufficient to explain today's high $CCR5$-$\Delta32$ frequency. Plague immediately shot to the top of our list of epidemics that may have catalyzed the remarkable rise of $CCR5$-$\Delta32$.

The hypothesis was alluring and made good biological sense. The idea also presented a few testable predictions. The plague bacillus, *Yersinia pestis*, actually causes death by injecting toxins into macrophages and lymphocytes expressing CCR5. Macrophages are the precise cells that HIV enters via the CCR5 receptor and the cells in which HIV-1 resides sequestered from the immune defense artillery. They are also one of the cell types that serve as reservoir where HIV can hide from powerful anti-HIV drugs. HIV also infects CD4- and CCR5-bearing T-lymphocytes in blood and in lymph nodes, again the same ones to explode as buboes when infected by the plague bacillus.

A curious prediction of the CCR5-plague connection would be that the normal CCR5 receptor would itself play a physiological role in bubonic plague disease as it does in AIDS. Are individuals with two copies of $CCR5$-$\Delta32$ resistant to plague as they are to HIV? Answering that question has proven difficult because it would be unethical simply to expose people to *Yersinia pestis*. Less than a thousand people worldwide inadvertently get exposed to *Yersinia* each year, mostly in Asia and India, places where $CCR5$-$\Delta32$ does not occur. We do not have an answer to this yet.

My trip to Eyam was actually prompted by Jennifer Beamish, an inquisitive television producer for London's Channel 4 (no relation to Douglas Leo Beamish), who wondered if the unusual circumstances of the plague years in that village might have led to a rise in $CCR5$-$\Delta32$ among the survivors and in their descendants. With Jennifer's help, we persuaded Eyam descendants of plague survivors to volunteer a cheek swab for $CCR5$ genotyping to test the prediction. We actually did find a modest increase of $CCR5$-$\Delta32$ allele frequency in Eyam, 15%, slightly higher than the 10% seen in other English villages. In addition, there were twice as many homozygous $CCR5$-$\Delta32$

people in Eyam than we found elsewhere. The results were encouraging, but not conclusive. The numbers were too low to be significant in a statistical sense.

Another idea we are considering is to look for a difference in the sensitivity of macrophages from people with the alternative *CCR5* genotype to *Yersinia* toxin in a laboratory test. We are also testing plague-sensitive mouse strains, some with normal *CCR5* genes and others with the *CCR5* gene "knocked out," inactivated by genetic engineering. If *CCR5-Δ32* was elevated by the plague in medieval times, then it should have a direct influence on *Yersinia's* ability to kill human and mouse cells. We await the data for a definitive answer.

Alternative explanations for the selection of *CCR5-Δ32* are also possible. Outbreaks of disease before historic times could have played a role. The 680 years ago date is a very rough estimate. It could be off by hundreds of years, and, more important, it simply represents the time for the most recent selective episode. If earlier selective elevations of *CCR5-Δ32* occurred, and I believe they did, then the mutation itself may be thousands of years old. Swedish researchers Kerstin Liden and Anders Gotherstrom recently discovered three *CCR5-+/Δ32* heterozygotes and one *CCR5-Δ32/Δ32* homozygote among skeletal remains of twelve people in Neolithic Scandinavian grave sites; they were dated at 3028–2141 B.C. If this is correct, the actual age of *CCR5-Δ32* is much older than the medieval plague that we impute elevated its frequency. Perhaps a prehistoric blight in northern Europe preceded the Justinian Plague. This scenario is attractive since it would explain the gradient of *CCR5-Δ32* frequency peaking in Scandinavia today.

A variety of other devastating human disease outbreaks might have catalyzed the rapid rise of *CCR5-Δ32*. Plausible candidates include tuberculosis, cholera, smallpox, anthrax, and measles. Each of these caused enormous epidemic mortalities in historic times in Western civilization. Two of these candidates—smallpox and measles—draw support from functional experiments. A close relative of human measles and smallpox viruses, rabbit myxoma pox virus actually utilizes CCR5 as a cell entry receptor like HIV-1. If the human smallpox virus also interacts with CCR5 on the cells it infects, that agent, which also killed millions in historic times, may also help explain the rise of $CCR5\text{-}\Delta32$. Testing the interaction of smallpox and CCR5 was nearly impossible since remaining smallpox stocks are vigorously guarded to prevent dissemination of a disease once thought to be nearly eliminated. That may change in the wake of recent bioterrorism concerns.

Historians and scientists piecing together the evidence to reconstruct our past speculate and interpret in a chaotic fashion, always modifying, tinkering, criticizing, and erecting new ideas. In the end, some inherent resistance to plague saved Margaret Blackwell and her kin. If it was her genetic endowment, whether $CCR5\text{-}\Delta32$ or another, the benefit was passed on to her offspring to act again on the next emergence. Sadly, there is always a next emergence.

During the plague years, the survivors of previous outbreaks always did better than those unexposed, a close-up consequence of natural selection exerted by the disease. Jewish peasants did seem to survive better and they paid a heavy price of persecution. Today, $CCR5\text{-}\Delta32$ has a remarkably high incidence (around 16%) in Israel, an exception to the 5% frequency distribution of southern Europe. Could it be that the plague history has favored a $CCR5\text{-}\Delta32$ rise in that community?

Our human gene script is just now beginning to reveal the footprints of long-past selective events, of disease outbreaks, and of adaptive episodes. Modern genomes carry the master notes for countless successes in a vast natural experiment. How fortunate that the subtle solutions we uncover, what makes us survive, can be harnessed with coming technology to frame a world and society vastly better equipped to ease human suffering than the vicars and prelates of the fourteenth century. This is the promise of the new era of postgenomic biotechnology.

Fourteen

A Silver Bullet

JESSE GELSINGER WAS DRIVEN BY HIS NATURAL EMPATHY. The Tucson teenager was excited for the opportunity to help the thousands of children who suffered from an obscure genetic malady he knew too well, OTC (ornithine transcarbamylase) deficiency. The rare disease occurs in about one in forty thousand births. It produces a defect in urea metabolism that results in elevated blood ammonia, organ toxicity, brain damage, and liver failure. Afflicted newborns fall into a coma at birth and die within a few months.

Jesse had the bad luck to be born with the disease, but mercifully a rather mild form, one in which only some of his body's cells were defective. His affliction was controlled by a very low protein diet and a daily regimen of thirty-two pills to cleanse his system of the threatening ammonia poison. His few adolescent tinkerings with neglecting doses had led to fierce stomach cramps and uncontrollable vomiting. After his dad discovered him retching on the sofa in toxic distress in December 1998, he never missed a dosage and avoided the joys of burgers and hot dogs completely.

Jesse's condition was diagnosed when he was two years old. By the time he was seventeen he understood there was no cure, only cumbersome symptomatic treatment. His pediatrician told him of a bold new clinical trial under way at the prestigious Institute for Human Gene Therapy (IHGT) at the University of Pennsylvania to test a treatment involving delivery of a normal OTC gene directly to the liver of genetically OTC-deficient children. The institute, headed by gene therapy pioneer James Wilson, had received U.S. Food and Drug Administration (FDA) and NIH Recombinant DNA Advisory Committee (RAC) approval to recruit relatively healthy but at-risk patients for a Phase I clinical trial. This first step in gaining FDA approval for new treatments is meant primarily to gauge the safety of the compound, not necessarily to discover whether it actually works. Determining a new drug's effectiveness would be the goal for Phase II clinical trials, typically involving a few hundred volunteers. The final stage, Phase III, to which less than a third of new pharmaceuticals progress, involves thousands of patients that suffer the target disease. Phase III clinical trials are designed to explore the drug's effectiveness, side effects, and benefits relative to available alternative treatments.

Wilson and Arthur Caplan, director of bioethics at Penn and a widely respected expert on patients' rights in medical trials, reasoned that OTC-affected newborns would be inappropriate for a Phase I trial since they could hardly give informed consent themselves, and their grief-stricken parents were unlikely to be able to provide an objective informed consent either, being haunted by a deadly disease in their new baby. They decided that healthy volunteers were preferable, at least for the Phase I safety test. Wilson had described the OTC gene therapy trial's goal as a search for the maximum tolerable dose—low enough to avoid adverse side effects, but high enough to reverse the symptoms.

Jesse was anxious to volunteer for the trial. "What could happen?" he quipped to his buddies. "If I die, it's for the babies." But he still had to wait until he was an adult.

On his eighteenth birthday, June 18, 1999, Jesse, his dad, his brother, and two sisters flew to Philadelphia for a sight-seeing holiday as well as to enroll Jesse in the Penn protocol. The family met with Dr. Steve Raper, surgeon and principal investigator of the experiment, who told Jesse that he qualified to receive the treatment. Jesse returned to the Institute for Human Gene Therapy by himself that fall. On September 13, he was infused with thirty-eight trillion particles of a genetically modified adenovirus, a human respiratory virus engineered to carry a normal human *OTC* gene. The altered gene therapy virus was injected through catheters into his groin leading to the hepatic artery, a direct shunt to his liver. Jesse received the largest dose in the trial; he was the youngest of eighteen volunteers.

Within a few hours of the treatment, his organ systems began to unravel. His stomach ached, and his fever spiked to 104.5°F. Liver bilirubin (the cause of jaundice) shot to four times normal and his blood-clotting apparatus shut down. Over the next few days Jesse fell into a coma, his kidneys failed, his lungs collapsed, and his brain waves flattened. An overwhelming inflammatory reaction to the adenovirus had caused a massive immune shock and multi-organ shutdown that would extinguish Jesse's young life four days after the infusion. It was the first mortality of a gene therapy patient in the struggling field's ten-year history. Jesse Gelsinger's death made front-page headlines in the nation's newspapers. The catastrophe rocked the very foundation of a promising discipline. It would hobble a blossoming institute and devastate Jesse's clinicians, his caregivers, and his family.

The promise of gene therapy was from its onset rather simple to grasp, to explain, and even to sell. First, identify a patient suffering from one of the thousand-odd hereditary diseases caused by an identifiable gene defect. Then splice a normal gene into a delivery vehicle or vector, usually a human virus that has been genetically disarmed by removal of a few necessary virus genes. Identify the organ that is lacking the gene product and deliver the vector carrying the normal gene to the tissue. Cross your fingers and hope that the tissue allows the new gene to be expressed, thereby reversing the genetic disease.

If it could be done, the payoffs would be tremendous. We could correct sickle-cell anemia or Tay-Sachs disease. Treat cystic fibrosis. As the genes that impede cancers are discovered, doctors may be able to dispatch a gene therapy "smart bomb" to knock out the tumors. One might cleanse HIV-1 or hepatitis viruses from the system with powerful "anti-sense" gene constructs that block virus gene action in its tracks. One day we may deliver cloned antibody genes to people so everyone could enjoy the most effective immune defenses ever seen regardless of their own inherited immunological endowment. Gene therapy could deliver good genes to diseased tissue and right the wrong. It could be the silver bullet of the dazzling biotechnology revolution.

The prospects for gene therapy are alluring. Hundreds of devastating hereditary diseases are incurable save for alleviating the symptoms. Genetic engineering was almost too easy to articulate, masquerading the extraordinary difficulties the field would encounter. One snafu after another—technical, political, ethical, and procedural—would dog the field over its brief heyday. Some highlights of the discipline's troubled history merit brief mention.

Gene therapy actually started out in the doghouse publicity-wise when in 1980, a UCLA hematologist, Martin Cline, chose to ignore NIH review board constraints and inject cloned hemoglobin genes into Israeli and Italian children suffering from a fatal blood disease, thalassemia. His application to test American children had been rejected by an ethics review board at his medical school as too much, too soon, too fast, and too risky. Cline was censured loudly for going ahead recklessly, in an overseas setting. He was forced to resign as chairman of his department at UCLA and heavily criticized for his arrogance in skirting ethical conventions for institutional review of medical trials.

The field simmered for a decade as W. French Anderson, a persistent and determined laboratory chief at NIH, slowly and deliberately maneuvered through a sea of bureaucratic, ethical, and medical reviews to gain the first legitimate approval for a gene therapy trial. His subject was a four-year-old child, Ashanti DeSilva, who suffered from a rare hereditary immune deficiency called "severe combined immune deficiency" (SCID). SCID results from a rare mutation in the gene for the enzyme adenosine deaminase (ADA). This defect causes a devastating breakdown in immune function, leaving its victims defenseless against common bacterial and fungal infections. The condition was nearly always fatal unless patients were housed in a sterile, hermetically sealed environment protected from outside microbial agents. The most famous SCID case was a Houston youngster, David Vetter, who lived in a sterile bubble-habitat until his adolescence. As a teenager, he chose to receive a bone marrow transplant hopefully to free him from his medical incarceration; but shortly thereafter, he succumbed to a massive fatal infection of Epstein-Barr virus, an agent that causes teenage mononucleosis and seldom kills people with intact immune systems.

By 1987, ADA purified from cattle tissues was shown to ameliorate SCID symptoms, but not very efficiently. The disease was an ideal candidate for biomedical science's first gene therapy attempt. It was a rather well understood single gene defect that affected lymphocytes, easily accessible for therapeutic manipulation as they circulated in the bloodstream.

Anderson and his NIH colleagues Michael Rosenberg and Kenneth Culver removed bone marrow lymphocytes from the little girl and injected them in with a normal *ADA* gene packaged in a retrovirus vector (a mouse virus with broad species infectivity including human lymphocytes). The transferred ADA began to function in the engineered cells and on September 13, 1990, nine years to the day before Jesse Gelsinger would receive his infused genes, Ashanti's genetically engineered cells were returned to her bloodstream.

The result was encouraging, but far from a true therapeutic success. Her cells made some ADA, but not a lot. Ashanti and about a dozen other genetically treated SCID patients are healthy today, but their engineered bone marrow cells did not make enough ADA or persist long enough in the bloodstream to risk terminating the bovine ADA supplemental treatment.

After Anderson's groundbreaking work, additional gene therapy protocols started to receive approval from the NIH-RAC committee, a watchdog group charged with review and regulation of genetic engineering clinical trials. Venture capitalists invested tens of millions in start-up biotechnology companies determined to harness the power of the gene therapy revolution. Over three hundred protocols and some four thousand patients with twenty different diseases—including SCID, cystic fibrosis, Gaucher's disease, AIDS, and cancer—were targeted for clinical gene transfer protocols throughout the 1990s. Modest signs of transgene expression (a transgene is a gene delivered to a cell or person by gene therapy engineering) and short-lived clinical improvements were heralded by an insatiable media enamored with the pioneering new medical procedures.

But as more and more gene therapy trials were undertaken, the sobering reality was sinking in: Real cures weren't being achieved. Expression of the transgenes was nearly always too low, or transient (it simply disappeared after infusion), or both. And scientists and doctors constantly worried about potential adverse side effects, particularly a devastating immune reaction, anaphylactic shock, or autoimmune destruction.

By the mid-1990s, NIH director Harold Varmus, himself a molecular biologist, commissioned a blue ribbon panel to evaluate the new field and all the hype that accompanied it. The panel was headed by the respected human geneticists Stuart Orkin and Arno Motulski and its December 1995 report was unabashedly frank in its criticism of weak science and overpublicizing of the field. The report, which Varmus readily endorsed, noted, "While the promise of gene therapy was great . . . clinical efficacy has not been definitively demonstrated" for any disease. It chided the champions of gene therapy for "overselling the results of clinical studies leading to a mistaken perception that it is further developed and more successful than it is" and for understating potential risks from immune hyperreaction and to cancers. The committee suggested that such exaggerated pronouncement of success or cures "threaten confidence in the integrity of the field." The authors called for better basic research of the physiological details of target diseases and for better vectors that could not only deliver genes effectively to specific tissues, but also control their expression. The panel also openly endorsed collaborative partnerships between academic investigations and commercial biotechnology industries to facilitate resource development suitable for clinical applications.

Varmus did not hide his disdain for the foundation of shoddy science on which he felt the entire gene therapy discipline was built. He believed the field was tainted by exaggerated success reports in popular media that appeared orchestrated to stimulate investment in the commercial biotechnology industry. He welcomed the caustic report and the tighter and stricter reviews of gene therapy protocols that followed.

James Wilson's IHGT at Penn escaped serious damage from the scathing report because of their proven excellence in gene therapy development and a reputation for cautious experimental design. Wilson commented that the Orkin report "helped us, because it eliminated a lot of hacks."

The experiences of gene therapy subsequently improved. A French team led by Alain Fischer reported in the spring of 2000 that they had achieved the first real success in treating three SCID-X1 patients with a transgene. SCID-X1 patients are immune suppressed like the SCID-ADA patients, but due to a mutational defect of a different gene, one that encodes the γc-cytokine receptor (a cell surface molecule that mediates immune cell signal communication). Untreated SCID-X1 newborns inevitably die of common microbial infections before their first birthday.

Blood cells from three infants with the disease, ages three, eight, and eleven months, were processed to purify a type of white blood cell called hematopoietic stem cells. *Hematopoietic* means "of the blood," and stem cells are immature precursor cells that have not yet developed into immune specialists such as T-lymphocytes, B-cells, or macrophages. Stem cells have the ability or potential to differentiate into these specialized cells and will do so in response to cellular signals sent out in lymph nodes, spleen, and other tissues. However, until they are so provoked, the stem cells divide forever, unlike their descendent differentiated cells, which do their specific task in the body and then are programmed to die. The stem cells' long life, even immortality, made them ideal vehicles for persistent transgene delivery and expression.

Fischer's team injected the infants' hematopoietic stem cells with a virus vector carrying a high-expressing copy of the healthy *γc-cytokine* receptor gene the children lacked. Each child's lymphocytes began to express the gene in their bloodstream. The treatment empowered their immune system to provide, in the authors' words, "full correction of disease phenotype and, hence, clinical benefit." The children got well and returned home. Some two years afterward, the children were miraculously leading apparently normal lives, adequately protected by their transgene-complemented immune system.

Two years later, eleven SCID children had been treated and nine were similarly cured.

The SCID-X1 result would seem to represent the first verified gene therapy to benefit an actual patient. These children would be closely monitored for years to come, but by mid-2002 the results were encouraging. The reason this treatment succeeded where others before have failed seems to lie in the details. The clinical trial relied on a number of technical modifications learned from a decade of experimental work in animal and human immunobiology. The team targeted hearty and long-lived stem cells for gene delivery, increasing the chance of transgene persistence in the transfused patients. Fischer's protocol also selected a genetic disease with a major single gene defect expressed in an accessible tissue. They chose a delivery vector (the mouse retrovirus) that would optimize gene expression in lymphoid cells of the immune system, the very cells they wished to target.

Stuart Orkin, chair of the critical 1995 NIH gene therapy report, complimented the SCID-X1 success, saying, "They did all the science beforehand and translated it to the patient, which is what the field should be doing."

There have been other new studies that are almost as encouraging. For example, hemophilia-B patients who lack a blood-clotting factor called factor IX, and hemophilia-A patients who lack Factor VIII, have been treated with transgenes that modestly supplemented the clotting factor but then disappeared within a few months. Still, these promising outcomes can be built upon in the future.

Transgenes have shown real potential for treating cancers, particularly prostate cancer and skin cancer. One particularly innovative approach to cancers involves a genetically engineered virus Onyx-015 developed by Frank McCormick, director of the University of California at San Francisco Cancer Center. The vector is a cell-killing adenovirus, but one with a disarmed virus gene, *E1B*, which is normally programmed to destroy the important $p53$ in cells it infects. The $p53$ gene is a lead player in a cell's security detail. In cells under attack by viruses, or with damaged DNA, the $p53$ gene detects the damage and prevents cell division. So Onyx-015 cannot grow in normal cells because their p53 senses the virus and stops the cells replicating.

However, cancerous cells very frequently have a $p53$ gene damaged by the cancerous event, so Onyx-015 virus will infect these and destroy them.

McCormick cleverly attempted to engineer a "smart bomb" virus that would attack and kill tumor cells but would not harm normal cells. In a Phase II clinical trial of thirty-seven patients, Faldo Khari and his collaborators at the University of Texas treated patients with head and neck cancers using a combination of Onyx-015 and conventional chemotherapy. The tumors dropped to half their original size in nineteen of thirty treatments (63%) and the tumor disappeared completely in eight of the patients treated with Onxy-015. By contrast, every patient in the control group who received chemotherapy alone showed increased tumor growth.

W. French Anderson, a pioneer in gene therapy research, applauded the research as a score for gene therapy against cancer. But Frank McCormick, the developer of Onyx-015, tried to distance his invention from the beleaguered field, commenting that the strategy was not gene therapy at all. His reasoning was that no new human genes were introduced, only a virus with a defective $E1B$ gene.

The tragic events at Penn's Institute for Human Gene Therapy cast a broad dark shadow over any jubilation researchers might have felt over their advances. How could such a catastrophe have happened? Volunteers had died in trials before, but nearly always from complications of their incurable disease. Not so here. Jesse Gelsinger was not seriously ill; his condition was under control. From all indications, his death was caused by the gene therapy itself. So what went so very wrong?

In the days and weeks following Jesse's death, IHGT researchers, University of Pennsylvania officials, and NIH and FDA regulators pored through the details of the OTC trial: patient recruitment, counseling, informed consent, protocol, and reviews, all in the fishbowl of conspicuous media scrutiny. Headlines were unforgiving and accusatory. Someone must be held accountable; a reckoning was inevitable.

The FDA announced the preliminary results of their inquiry in December 1999 and put the blame squarely on the Penn team who designed, implemented, and endorsed the protocol. In a terse eighteen-point critique, the FDA charged that Gelsinger was ineligible for the study protocol's criteria because his liver was not well enough. His blood showed unacceptably high ammonia levels, elevated 30–60% before the transfusion, at the beginning of the trial. His bone marrow was also depleted of lifesaving erythroid precursor blood cells.

The FDA report also faulted the Penn researchers for their failure to report serious adverse side effects associated with the adenovirus vector in treatment of four earlier patients. They charged that the informed consent form that Jesse signed was misleading and inadequate because it made no mention of the adverse patient reactions or the deaths of monkeys similarly treated with adenovirus in experimental settings. The monkeys had received a higher dose than the Penn trial volunteers, about twentyfold more, but one of the monkeys developed the same rare blood disease that afflicted Jesse.

Additional warning signs of trouble with the adenovirus vector surfaced with the benefit of 20/20 hindsight. Jesse's autopsy revealed that the virus had disseminated not only to his liver but also equally effectively to most of his other organs (spleen, lung, thymus, heart, kidney, testicles, brain, pancreas, lymph nodes, bone marrow, bladder, small intestine, and skin). This was an unwelcome complication for a vector supposedly delivered directly to the liver through the hepatic artery. Team member Mark Batshaw had suggested earlier that the virus carried a "zip code" on it that would deliver and restrict it to the liver. The zip code failed in Jesse's treatment.

Related adenoviruses were also well known to cause fatal liver failure in dogs and birds, and to disseminate widely in multiple tissues in horses. To make things even more troubling, Wilson and the University of Pennsylvania had business entanglements with a biotechnology company, Genova, Inc., that stood to profit from their gene therapy protocol. Wilson had started the company, which provided 20% of the support for Wilson's laboratory at the time of Jesse's death. Alan Milstein, Jesse's family lawyer, charged that patents and businesses held by Wilson, by the university, and by IHGT clouded their judgment in conducting the clinical trial.

The FDA suspended IHGT's permission to conduct clinical trials on human patients in January, and by May the University of Pennsylvania suspended indefinitely the institute's ability to conduct any human clinical trials. Jesse's father, Paul Gelsinger, filed a civil lawsuit, citing as defendants the University of Pennsylvania, IHGT investigators and clinicians, and William Kelly, dean of Penn Medical School and holder of some Genova patents. Also named initially (but later dropped) as a codefendant was Arthur Caplan, University of Pennsylvania bioethicist, for collaborating in the decision to recruit healthy volunteers and also for his role in preparing the "misleading" informed consent form. It was the first time an ethics counselor was named in a clinical trial legal action alleging ethical fault.

The university settled the complaint quietly in November 2000. The exact amount paid to Paul Gelsinger is confidential but observers close to the case estimate the settlement figure between five and ten million dollars. In December 2000, the FDA proposed that James Wilson be officially "disqualified" as a clinical investigator for "repeatedly and deliberately violating regulations governing the proper conduct of clinical studies." Disqualification is as severe a sanction as can be levied against a clinical researcher short of pulling his medical license.

In the aftermath of Jesse Gelsinger's death, some important messages came through loud and clear. Gene therapy protocols simply do not yet have ideal tools for successful treatment. The gene delivery vehicles are not tissue-specific as advertised, and they fail to precisely adjust the amount of delivered gene product properly when they do find the right tissue. Further, most vectors fail to persist. One day, gene therapists will be able to consult a reference text or Web site that lists explicit properties of experimentally tested gene promoters, gene enhancers, and gene operators. These are the names we give to the mysterious group of gene regulatory sequence elements in our genomes that determine in which tissue and to what extent a gene is expressed, as well as how long the vector or gene product persists.

But the science is not there yet. The truth is that much more basic research is required before we can really predict and control precisely the action and genes we deliver.

The problem is comparable to repairing or doing "therapy" on an automobile with defective brakes. Without a manufacturer's auto repair manual specific to the make, model, and year, any "treatment" that actually repairs the brakes would involve unusual good fortune. Could we hurl a set of new brake pads at a car with bad brakes, ignorant of where they connect, the proper brake fluid levels, the connections to the foot pedal, and precision brake adjustments, and hope for success? How often would that treatment result in a car with dependable brakes? Would you drive such a car onto a freeway?

The 1995 NIH report warned that the field needed more basic research development on the diseases, on the details of the defects, and on the intricacies of fine-tuning gene action in a clinical situation. Perhaps because the concept of gene therapy is so straightforward, so easy to explain, and so full of promise, its practitioners suffered from false confidence around the technical but critically important minutae. For gene therapy, now more than ever, the devil is in the details.

This should come as no surprise if we consider the learning curve for other therapies. Theodore Friedman, one of the pioneers of this young field, penned a frank editorial in *Science* reminding his colleagues to learn from the trials and tribulations of vaccine and drug development. Research and development phases for these are frequently tedious, expensive, and for the most part dead-ended. Gene therapy will be no different, he argued; it is simply younger. Friedman writes:

> *The pharmaceutical industry, more mature and experienced than the gene therapy community, devotes enormous research and financial resources to studies of the biodistribution, pharmacological properties, stability, and metabolic properties of a potential new drug as well as the physiological, immunological toxic effects on the host. Despite such care, because of the enormous complexity of human physiology and disease, and because even the most extensive animal data do not always faithfully predict responses in humans, adverse clinical responses have*

occurred and will again. Some clinical applications [of gene therapy] have simply outstripped scientific understanding of the disease model or the properties of the vectors, resembling an army too far ahead of its supply lines.

Veteran *New York Times* correspondent Nicholas Wade adds a pithy summation to the concept. "Evolution's programming took 3.5 billion years to develop; learning to fix the bugs in it is not going to happen overnight."

An unspoken concern is that the lengthy bureaucracy of clinical trial ethical reviews themselves may have blunted the vigilance of the Penn researchers. French Anderson spent a decade of tedious preparation and ethical reviews to get permission to attempt gene therapy with Ashanti DeSilva. Wilson's IHGT was the premier center of the field, but still, getting permission to try eighteen patients was a long, involved bureaucratic exercise. Suppose ethical constraints had not limited the trial to eighteen subjects, but allowed a larger group of, say, five hundred. That higher number of patients would have allowed an incremental step-wise Phase I screen for adverse reactions. Could it be that the extreme scrutiny itself led to careless judgment that might have been avoided in a much larger trial?

Even more disturbing, the tortured politics of the gene therapy field itself may have contributed to lightened oversight of safety issues. In the wake of the Orkin report, NIH director Harold Varmus slashed the standing NIH-RAC committee from twenty-five to fifteen members and stripped it of approval authority. He viewed the NIH-RAC approval process as enabling to the public relations hyperbole driven by the biotechnology industry. However, his strategy backfired because undercutting the NIH-RAC's authority and responsibilities left the FDA as the sole arbiter of gene therapy protocols. By law all FDA deliberations on clinical trials are secret, while the NIH-RAC discussion had been public, very public. A *Nature* editorial in June 2000 faulted Varmus and NIH's leadership for diminishing the NIH-RAC resolve, perhaps contributing to a sense among researchers that reporting of adverse effects was not compulsory. In the end, plenty of safety lapses and armchair finger-pointing emerged from the crisis, leaving scientists scrambling to fix the tainted image of the promising young clinical discipline.

In August 2002, the gene therapy community received yet another jolt. One of Alain Fischer's eleven SCID children, each treated with a normal γc-cytokine gene receptor gene, developed leukemia. The mouse retroviral vector carrying the transgene integrated into a lymphocyte chromosome of the three-year-old adjacent to a well-known oncogene, *LMO2*. The retrovirus switched on the oncogene's expression, unleashing cell proliferation and leukemia. Then in January 2003, another three-year-old child in the same trial developed leukemia, and with the retroviral vector lodged in the same *LMO2* oncogene. By February, a third child in the French study showed evidence of the vector integration in the same deadly spot.

Fischer's trial was halted immediately. The patients were hospitalized and placed on powerful cancer chemotherapy. The U.S. FDA and several other countries quickly halted all SCID and retrovirus-mediated gene therapy trials. The worst fears of the 1995 NIH report on gene therapy had come true: cancer produced by a leukemia virus vector that chose to integrate in the wrong place, three times. It was thought to be extremely unlikely, but it happened to the French children. And supposedly disarmed retroviral vectors derived from leukemia viruses were the preferred delivery vehicles for over 75% of all gene therapy trials. The only real success for human gene therapy had stumbled badly.

The post-genomics era, the one we entered once the human genome sequence was drafted, will surely see a renaissance of better, more effective, and safer gene therapies. By the close of the twenty-first century, scores if not hundreds of human hereditary syndromes will be treated by gene-based correction. The therapeutic proof of principle was established by Fischer's now tempered success with the *SCID-X1* children and in a dozen genetic disease corrections in mouse models. Identification of single nucleotide variants that mediate complex human diseases like diabetes, multiple sclerosis, and hypertension will grow to number in the thousands. Patients' genotypes at critical loci will soon become as important as their clinical symptoms in both diagnosis and treatment regimens.

Before that happens, scientists need to better grasp the basic workings of our genes, their regulatory elements, and intercellular molecular communications. The subtleties of each disease and the human body's repair systems must be detailed as if we were dissecting an enemy nation's armed forces: reveal the inner workings of each battalion, each division, each squadron—the infantry, artillery, marine, and air capabilities—the entire militia. Pathology and immunobiology will get there one day, but it will take time, money, dedication, and considerable research talent. Fulfilling the dreams of genetic engineering requires a more thorough understanding of normal human development, from egg to senescence, as well as specific disease pathogenesis, and adverse events—how things can go terribly wrong. A far more comprehensive perspective on disease breakdown and on working physiological processes are critical for designing and evaluating treatments that will be both safe and effective.

The strategies for gaining this deep understanding of biological goings-on have been the underlying theme of this book. Harness the lessons from a vast and extensive genomic record, encrypted in the genes of living mammals, each connected to humankind by deep evolutionary roots. Mine their genomes for evolved adaptive counterbalances to their diseases, and develop a parallel gene-based treatment to homologous human maladies. Quietly selected, time-proven, and exquisitely fine-tuned gene modifications, evolved to abrogate chronic cancers, physiological degenerations, and hereditary syndromes, are lurking in living species whose ancestors have won their own evolutionary battles.

California Institute of Technology President David Baltimore coined the term *intracellular immunization* in 1988 to describe the concept of delivering virus-inhibiting restriction genes to infected patients as a possible gene therapy for virus diseases like AIDS, hepatitis, or Ebola. Attempts to apply this concept have so far been flawed because, to be honest, we are not really sure which genes to try. Nor have we optimized a system to target those genes to a specific tissue or to fine-tune their expression. Yet, the wild California mouse of the first chapter overcame a lethal epidemic using just such a technique. The mouse's Chinese great-great-grandparents acquired a foreshortened version of the virus, turned it on, and regulated its expression just right in exactly the cells the fatal virus destroys. That accomplished, the new gene was transferred to their offspring as a natural intracellular immunization, a cure to a fatal infectious disease. Similar solutions may be lurking in the genome of wild monkeys, big cats, giraffes, woodchucks, or armadillos. Molecular biologists and disease gene hunters need only to befriend a field biologist, a naturalist, and a wildlife veterinarian.

There are lessons as well in our own history. A benevolent human gene mutation, *CCR5-Δ32*, provides resistance to its carriers against AIDS. The protective *CCR5-Δ32* allele was born in historic times and won itself rapid elevation in frequency by providing protection to its carriers during waves of the Black Death in fourteenth-century Europe. The major histocompatibility complex (MHC) of humans and animals contains a garden of evolutionary secrets, a majesty of genomic adaptations all spawned by the need to recognize the myriad infectious scourges that hammer free-living species like ourselves. Medical science is just beginning to explore ways to engineer the MHC to improve our immune response.

We must remember that endogenous virus elements, inactivated pseudogenes, and the delicate interplay of hereditary defect and infectious disease avoidance are here for a reason. Hereditary diseases such as sickle-cell anemia, cystic fibrosis, and Tay-Sachs disease remind us of nature's silent balancing act. The sickle-cell defect once protected its carriers against malaria, the cystic fibrosis mutation against typhoid, and Tay-Sachs disease possibly against tuberculosis. But their descendants pay a price. Our ancestors inadvertently mortgaged the coming generations with hereditary handicaps that in another time conveyed avoidance of their own deadly pestilence.

Not many of these genetic secrets would have been revealed tinkering with lab mice. Instead they have come from population scans of humans and a vast array of other mammals. The tools for gene therapy, first developed in laboratory mice, have now been extended for gene delivery technologies in rats, monkeys, dogs, cats, sheep, and pigs. The comparative insight domestic species will provide has only just begun to accumulate. The world's veterinary schools promise to become centers for broad-based medical advances in the coming years. And beyond domesticated animals, the world of free-ranging species offers genomic stories that consistently point to new proven natural defenses against hereditary, infectious, and neoplastic disease.

The disciplines of molecular biology and genomics are at last in a position to accept these tremendous gifts. Nature has composed and performed countless experiments, replete with written notations that extend our perspective in three very provocative dimensions. First is the intricate human genome script orchestrating virtually every aspect of human biology. Second is the comparative genomic insight derived from sampling parallel development, specialization, and adaptation of the world's five thousand species of mammals. The third and deepest derives from the reconstruction of the genomes, the forms, and the functioning of the ancient ancestors of living mammals. The triangle sets out the key to an in-depth appreciation of the process, the mode, and the tempo of genomic changes that preceded modern genome organization and function. There was probably never a better time to pursue biological questions. We are at last beginning to unveil the mysteries of our genomic roots.

A generation ago, genomics didn't exist and the biotechnology industry was in its infancy. Engineers applied the laws of physics to architecture, to electronics, and to chemical processing. In the generations to come, engineering will occupy a new place in biomedical science. Gene delivery, cloning, and modification have already revolutionized crop agriculture to eliminate rotting, promote growth, and resist blights. Bioengineering institutes will one day stand beside traditional engineering schools on university campuses. NIH's latest initiative, the National Institute of Biomedical Imaging and Bioengineering, offers a preview to similar centers in public, university, and corporate environments where practitioners of genetic biotechnology will develop customized biomedical and futuristic applications.

Countless other aspects of life will improve, but the challenge will be to move forward in a responsible and beneficial way. I hope our new insight will empower and challenge coming generations to stabilize the earth's biodiversity and in the process improve the health of humankind as well as that of the animal companions with whom we share our past, present, and future. It is a time for hope and for pause.

Glossary

Allele—One of several different forms of a gene that differ usually by a single DNA letter difference. This slight difference can cause changes in the end product of a gene function such as eye color differences, the presence of a hereditary disease, or variations in appearance.

Allozyme—An allelic enzyme, a common genetic variant in an enzyme caused by a single letter variation in the enzyme's coding gene. Allozymes are visualized by gel electrophoresis.

Analogous characters—Visible characteristics of different animals that look similar on the surface, but their developmental basis is distinct because they evolved at separate times on independent evolutionary lineages. Wings on insects, on birds, and on bats are analogous: same function—flap to fly—but of independent origins.

Antibody—A protein complex produced by an individual's immune system to defend against invading bacteria or viruses.

Antigen—A protein that stimulates an immune-mediated production of antibody.

Chemokines—Short proteins, 100 to 125 amino acids long, that are released by tissues damaged by abrasions, bruises, or infections.

Clade—A group of related species or DNA sequences that cluster together in a phylogenetic analysis.

Deletion—Mutational variant where a piece of DNA in a gene is simply missing; for example, *CCR5*-Δ32 is a deletion of thirty-two nucleotide letters from the normal *CCR5* gene.

Diploid state—The status of all mammals whereby each individual has two copies of each gene, one from each parent.

DNA—Deoxyribonucleic acid. The double helix genetic material that comprises our genomes, our chromosomes, and our genes; it comes in four varieties of nucleotides letters or base pairs: A (adenosine), T (thiamine), C (cytocine), or G (guanosine).

Ecotropic virus—A virus that grows in cultured cells of the same species from which the virus was found; for example, a mouse virus that grows in mouse cells but not in human or cat cells.

Electrophoresis—A laboratory technique whereby DNA or protein extracts from blood, cells, or tissue are placed in a mild electrical field on a Jell-O–like medium to cause their migration and separation. This method resolves genetic variants of proteins or enzymes (*see* allozyme) or DNA fragments (*see* RFLP).

Endogenous virus—A virus that lives in the host chromosome DNA and is passed vertically to offspring, in contrast to exogenous viruses, like flu or smallpox, which spread horizontally between individuals.

Exogenous virus—A virus that is transmitted among individuals as an infectious agent.

Exon—A part of a gene containing coding sequence of DNA.

FeLV—Feline leukemia virus. A retrovirus in domestic cats that causes leukemia and lymphoma.

Fibroblasts—A cell type found in skin pieces that can be planted in tissue culture medium and grown into immortal cell lines useful for DNA extraction and also chromosome analysis.

FIV—Feline immunodeficiency virus. The cause of AIDS in domestic cats.

Gene—A unit of information in DNA that specifies the translation (synthesis and assembly) of a particular protein. Human and other mammals contain around thirty-five thousand distinct genes in their genome.

Genome—A full-length copy of an individual's genetic endowment. A man's genome is the sum total of all his genes, his DNA, and his genetic information, neatly compiled in two distinct copies, one from each of his parents, in his every cell.

Genotype—The sum of two alleles, one from each parent, that an individual has at each gene locus.

Heterozygous—Refers to the state of a gene or locus in an individual when two different alleles are present, one inherited from either parent.

HIV—Human immunodeficiency virus. The cause of AIDS in humans.

Homologous characters—Visible traits on two different species inherited from a common ancestor; when examined in detail they usually show multiple complex and detailed similarities reflecting their ancestry and modification by evolutionary processes.

Homologous genes—Genes in different species that are descended from a common precursor of an ancestral species—for example, hemoglobin gene in humans and chimps and mouse or insulin gene in dogs, bears, and cats.

Homozygous—Refers to the state of a gene or locus in an individual when both parents contributed an identical allele type; a recessive disease or trait must be homozygous to be apparent.

Inbred mouse strain—A mouse strain derived from twenty or more generations of brother-sister incestuous matings; such intense inbreeding sheds population/strain genetic diversity by several hundred fold.

Intron—Intervening DNA between coding exons of genes.

Karyotype—Appearance of metaphase chromosomes of an individual; usually has a characteristic banded pattern among different species.

Lentivirus—A family of retroviruses that includes immunodeficiency viruses like HIV, FIV, and SIV.

Locus—A specific site or address in a chromosomal stretch of DNA, often including a gene or repeat sequence such as a microsatellite locus.

Lymphocytes—White blood cells that are responsible for mounting an immune defense against infectious viruses or bacteria.

Macrophage—A circulating blood cell specialized in defense against invading bacteria.

MHC—Major histocompatibility complex. A cluster of some 225 genes that occur together along a short chromosomal segment in the DNA of humans, mice, cats, and other mammals. About a dozen of the MHC genes encode proteins that coat the surface of cells where they engulf small peptides (short stretches of amino acids) from invading viruses as a prelude to immune-mediated demolition. Most of the MHC genes are extremely variable; some have over two hundred different alleles in outbred populations of mammalian species.

Microsatellites—Short stutterlike sequences found in chromosomes where a pair, a trio, or a quartet of nucleotide letters are repeated in tandem at least a dozen times. A microsatellite locus is a repeat stretch found at a specific chromosome site. All the mammals examined so far possess between 100,000 and 200,000 microsatellite loci (more than one locus) distributed in a nearly random manner across the entire genome. Because of the extraordinarily high variability among people, the 100,000 randomly spaced microsatellites have become a favorite marker for mapping genes because they are easy to follow in family studies. Microsatellites have also proved to be powerful tools for the forensic community in matching blood or semen samples left at crime scenes.

Minisatellites—Twenty-to-sixty-nucleotide-letter repeat sequences that produce a bar code pattern useful for individual recognition; originally the major tool of forensic DNA fingerprints.

Mitochondria—A cell's powerhouse where energy molecules are manufactured by combining atmospheric oxygen with a breakdown product of carbohydrate nutrients. All plant and animal mitochondria are themselves descended from a 600-million-year-old bacterial infection of early one-cell organisms. Today's mitochondrial DNA carries the remnants of that primitive invading bacteria's genes and uses them to make energy-rich molecules. Mitochondrial chromosomes are carried outside the nucleus in mitochondrial structures.

Molecular clock hypothesis—The premise that when the population of a species splits apart, perhaps by migrating across a huge river or mountain range, the descendants of the split-up populations would gradually change over time by acquisition of new mutations in their DNA sequences. As time passed, more and more mutations randomly dispersed across all their DNA stretches would accumulate. The longer the time elapsed, the greater the gene sequence divergence, so the amount of DNA sequence difference between two species is proportionate to the time elapsed since they split apart.

Murine—Pertaining to mouse, as in murine leukemia virus.

Neutral variation—Genetic variation that is not the object of natural selection pressure, and is free to drift about in a population without a strong adaptive advantage or disadvantage.

Nucleotides—The DNA letters (also called base pairs) of the genetic code that string together the genes in a chromosome. Single nucleotide letters or base pairs come in four basic varieties: adenosine-A, cytocine-C, thiamine-T, and guanosine-G. A string of nearly three billion nucleotides compose the genome of humans, cats, and other mammal species.

Oncogene—A small group of genes in human and other mammals that when mutated cause chaotic cell division and cancers; responsible for many inherited and spontaneous cancers.

PCR—Polymerase chain reaction. An enzymatic DNA photocopy technique, to make synthetic copies of genes extracted from tiny amounts of DNA, particularly useful for reading the gene sequences of trace amounts of DNA such as in forensic materials, plucked hairs, or other tissues.

Peptide—A short protein of fewer than twenty amino acids.

Phylogeny—A branching treelike diagram that connects different species to others based on the similarity of their homologous genes; phylogenies are meant to recapitulate the branching events in the evolutionary history of living species.

Plasma—The liquid part of blood that carries antibodies against infectious agents; differs from serum by the addition of an anticoagulant chemical such as heparin.

Polymorphism—A genetically variable site or locus in a genome, sometimes within a gene but more often outside genes in noncoding DNA.

Population bottleneck—An event that reduces the number of individuals in an outbred population to a small number, often for several generations, leading to a net reduction in overall genetic variability of the population.

Receptor—A large cell surface protein that serves as a docking station for a virus (a virus receptor) and allows its entry, or a recognition dock for an outside signal protein or hormone that stimulates the cell to react in some physiological function.

Recessive—An allele or disease that must be present in two copies in an individual, one from each parent, to be expressed or observed.

Retrovirus—A nasty type of virus that causes cancers, notably leukemias and lymphomas in chickens, cats, and mice. Retroviruses are unusual in that their genes consist of RNA—nucleic acids that control cellular activity—rather than standard DNA. These viruses use an enzyme to copy their RNA genetic code into DNA form, which it then inserts into its victim's DNA. RNA is usually a product of DNA, not the other way around, hence the prefix *retro*.

RFLP—Restriction fragment length polymorphism. A procedure that tracks DNA sequence differences in genes by the presence or absence of a specific DNA sequence that a bacteria-derived restriction enzyme will recognize.

RNA—Ribonucleic acid. A long counterpart molecule to DNA synthesized by enzymes in cell nuclei as an exact replica of expressed genes, transported to the protein synthesis machinery in the cytoplasm, and "translated" into a protein amino acid sequence through the genetic code.

SIV—Simian immunodeficiency virus. The virus most closely related to human HIV and found in twenty free-ranging African monkey species.

Subspecies—A population of animals that are geographically iso-lated, which over time leads them to become genetically and visually distinct from other populations of the same species. Siberian and Bengal tigers, or Florida panthers and Texas cougars, are familiar subspecies designations.

Taxonomy—The science-based hierarchical classification of the world's species. This includes not only the assignment of Latin names to distinct species, but also the sorting of species into genera, genera into families, families into orders, and so on. Also called "systematics."

Viremia—The state of being infected with a virus and showing large quantities of virus in the bloodstream.

Virogenes—Latent endogenous viruses nested in chromosomes that sometimes can become expressed and form whole virus particles.

Western blot electrophoresis—A procedure to detect antibodies in serum or plasma against a particular virus to which they were exposed. Disrupted proteins from purified virus are incubated with serum from an exposed individual. If infected or previously exposed, antibodies in the serum would bind to virus proteins. The complex of antibodies could then be separated and easily visualized on a Western Blot electrophoretic gel.

Suggestions for Further Reading

CHAPTER 1. A MOUSE THAT ROARED

Coffin, J. M., S. H. Hughes, and H. E. Varmus. 1997. *Retroviruses*. Plainview, NY: Cold Spring Harbor Laboratory Press.

Gardner, M. B., et al. 1991. The Lake Casitas wild mouse: Evolving genetic resistance to retroviral disease. *Trends Genet* 7:22–27.

Levy, J. A. 1993, 1994. *The Retroviridae*. 3 vols. New York: Plenum Press.

Radesky, P. 1991. A mouse tale. *Discover* (November), pp. 34–38.

CHAPTER 2. TEARS OF THE CHEETAH

Caro, T. M. 1994. *Cheetahs of the Serengeti Plains*. Chicago: The University of Chicago Press.

Hoelzel, A. R., et al. 1993. Elephant seal genetic variation and the use of simulation models to investigate historical population bottlenecks. *J Hered* 84:443–449.

O'Brien, S. J. 1994. A role for molecular genetics in biological conservation. *Proc Natl Acad Sci USA* 91:5748–5755.

O'Brien, S. J. 1998. Intersection of population genetics and species conservation: The cheetah's dilemma. *Evol Biol* 30:79–91.

O'Brien, S. J., et al. 1985. Genetic basis for species vulnerability in the cheetah. *Science* 227:1428–1434.

O'Brien, S. J., et al. 1986. The cheetah in genetic peril. *Sci Am* 254:84–92.

CHAPTER 3. PRIDES AND PREJUDICE

Driscoll, C. A., et al. 2002. Genomic microsatellites as evolutionary chronometers: A test in wild cats. *Genome Res* 12:414–423.

O'Brien, S. J. 1994. Genetic and phylogenetic analyses of endangered species. *Ann Rev Genet* 28:467–489.

O'Brien, S. J., et al. 1987. Evidence for African origins of founders of the Asiatic lion species survival plan. *Zoo Biol* 6:99–116.

Packer, C. 1992. Captives in the wild. *National Geographic Magazine* 181:122–136.

Packer, C. 1994. *Into Africa*. Chicago: University of Chicago Press, p. 277.

Packer, C., et al. 1991. Kinship, cooperation and inbreeding in African lions: A molecular genetic analysis. *Nature* 351:562–565.

Schaller, G. B. 1972. *The Serengeti lion—A study of predator-prey relations*. Chicago: University of Chicago Press, p. 480.

Sinclair, A. R. E., and P. Arecese, eds. 1995. *Serengeti II: Dynamics, management and conservation of an ecosystem*. Chicago: University of Chicago Press.

Sinclair, A. R. E., and M. Norton-Griffiths, eds. 1979. *Serengeti: Dynamics of an ecosystem*. Chicago: University of Chicago Press.

Wildt, D. E., et al. 1987. Reproductive and genetic consequences of founding isolated lion populations. *Nature* 329: 328–331.

CHAPTER 4. A RUN FOR ITS LIFE— THE FLORIDA PANTHER

Alvarez, K. 1993. *Twilight of the panther: Biology, bureaucracy, and failure in an endangered species program*. Sarasota, Fla.: Myakka River Press, p. 501.

Fergus, C. 1996. *Swamp screamer: At large with the Florida panther*. New York: North Point Press, p. 209.

Maehr, D. S. 1997. *The Florida panther: Life and death of a vanishing carnivore*. Washington, D.C.: Island Press, p. 261.

Roelke, M. E., et al. 1993. The consequences of demographic reduction and genetic depletion in the endangered Florida panther. *Curr Biol* 3:340–350.

Chapter 5. Bureaucratic Mischief

Federal Register 61: 26. February 7, 1996. Endangered and threatened wildlife and plants: Proposed policy and proposed rule on the treatment of intercrosses and intercross progeny (the issue of "hybridization") request for public comment.

O'Brien, S. J., and E. Mayr. 1991. Bureaucratic mischief: Recognizing endangered species and subspecies. *Science* 251: 1187–1188.

O'Brien, S. J., et al. 1990. Genetic introgression within the Florida panther *Felis concolor coryi*. *Natl Geo Res* 6:485–494.

Stevens, W. 1991. U.S. reviewing policy on protecting hybrids. *New York Times*, March 12.

Chapter 6. A Whale of a Tale

Baker, C. S., and S. R. Palumbi. 1994. Which whales are hunted: A molecular genetic approach to monitoring whaling. *Science* 265:1538–1539.

Baker, C. S., et al. 1990. The influence of seasonal migration on geographic distribution of mitochondrial DNA haplotypes in humpback whales. *Nature* 344:238–240.

Baker, C. S., et al. 1993. Abundant mitochondrial DNA variation and world-wide population structure in humpback whales. *Proc Natl Acad Sci USA* 90:8239–8243.

Baker, C. S., et al. 1996. Molecular genetic identification of whale and dolphin products from commercial markets in Korea and Japan. *Mol Ecol* 5:671–685.

Roman, J. 2002. Fishing for evidence. *Audubon* (January–February): 54–61.

Chapter 7. The Lion Plague

Brown, E. W., et al. 1994. A lion lentivirus related to feline immuno-deficiency virus: Epidemiologic and phylogenetic aspects. *J Virol* 68:5953–5968.

Carpenter, M. A., and S. J. O'Brien. 1995. Coadaptation and immuno-deficiency virus: Lessons from the Felidae. *Curr Opin Genet Devel* 5:739–745.

Morell, V. 1995. FIV—The killer cat virus that doesn't kill cats. *Discover* (July): 62–69.

Roelke-Parker, M. E., et al. 1996. A canine distemper virus epidemic in Serengeti lions (*Panthera leo*). *Nature* 379:441–445.

Chapter 8. The Wild Man of Borneo

Galdikas, B. M. 1995. *Reflections of Eden: My life with the orangu-tans of Borneo*. Boston: Little, Brown, p. 408.

Janczewski, D. N., et al. 1990. Molecular genetic divergence of orangutan (*Pongo pygmaeus*) subspecies based on isozyme and two-dimensional gel electrophoresis. *J Hered* 81:375–387.

Karesh, W. B. 2000. *Appointments at the ends of the world: Memoirs of a wildlife veterinarian*. New York: Warner Books, p. 379.

Lu, Z., et al. 1996. Genomic differentiation among natural populations of orangutan (*Pongo pygmaeus*). *Curr Biol* 6: 1326–1336.

Chapter 9. The Panda's Roots

Lu, Z., et al. 2001. Patterns of genetic diversity in remaining giant panda populations. *Conservation Biol* 15:1596–1607.

O'Brien, S. J. 1987. The ancestry of the giant panda. *Sci Am* 257:102–107.

O'Brien, S. J., et al. 1994. Pandas, people and policy. *Nature* 369:179–180.

Schaller, G. B. 1993. *The last panda*. Chicago: University of Chicago Press.

CHAPTER 10. THE WAY WE WERE

Lander, E., et al. 2001. Initial sequencing and analysis of the human genome. *Nature* 409: 860–921.

O'Brien, S. J., et al. 1999. The promise of comparative genomics in mammals. *Science* 286:458–481.

O'Brien, S. J., et al. 2001. Perspective: On choosing mammalian genomes for sequencing. *Science* 292:2264–2266.

Rettie, J. C. 1951. Forever the land: The most amazing movie ever made. *Coronet* 29:21–24.

Ridley, M. 1999. *Genome: The autobiography of a species in 23 chapters*. New York: Perennial/HarperCollins, p. 344.

CHAPTER 11. SNOWBALL'S CHANCE: GENOMIC PAW-PRINTS

Menotti-Raymond, M., et al. 1997. Genetic individualization of domestic cats using feline STR loci for forensic analysis. *J Forensics Sci* 42:1039–1051.

Menotti-Raymond, M., et al. 1997. Pet cat hair implicates murder suspect. *Nature* 386:774.

Reilly, P. R. 2000. *Abraham Lincoln's DNA and other adventures in genetics*. New York: Cold Spring Harbor Laboratory Press, p. 339.

CHAPTER 12. GENETIC GUARDIANS

O'Brien, S. J. 1998. AIDS: A role for host genes. *Hosp Pract* 33:53–79.

O'Brien, S. J., and M. Dean. 1997. In search of AIDS-resistance genes. *Sci Am* 277:44–51.

O'Brien, S. J., et al. 2000. Polygenic and multifactorial disease gene association in man: Lessons from AIDS. *Ann Rev Genet* 34: 563–591.

O'Brien, S. J., and J. Moore. 2000. The effect of genetic variation in chemokines and their receptors on HIV transmission and progression to AIDS. *Immunol Rev* 177: 99–111.

Radetsky, P. 1997. Immune to a plague. *Discover* (June): 61–66.

CHAPTER 13. ORIGINS

Carrington, M., et al. 1999. Genetics of HIV-1 infection: Chemokine receptor CCR5 polymorphism and its consequences. *Hum Mol Genet* 8:1939–1945.

Hahn, B. H., et al. 2000. AIDS as a zoonosis: Scientific and public health implications. *Science* 287:607–614.

Korbor, B., et al. 2002. Timing the ancestor of the HIV-a pandemic strains. *Science* 288:1789–1796.

Stephens, J. C., et al. 1998. Dating the origin of the *CCR5-Δ32* AIDS resistance gene allele by the coalescence of haplotypes. *Am J Hum Genet* 62:1507–1515.

CHAPTER 14. A SILVER BULLET

Friedman, T. 2000. Principles for human gene therapy studies. *Science* 287:2163–2165.

Gura, T. 2001. After a setback, gene therapy progresses . . . gingerly. *Science* 291:1667–1692.

Hacein-Bey-Abina, S., et al. 2002. Sustained correction of X-linked severe combined immunodeficiency by ex vivo gene therapy. *N Engl J Med* 346:1185–1193.

Stolberg, S. G. 1999. The biotech death of Jesse Gelsinger. *New York Times*, November 28.

Washburn, J. Informed Consent 2001. *Washington Post Magazine*, December 30, p. 8.

Acknowledgments

THERE ARE SO MANY PEOPLE WHO HAVE CONTRIBUTED TO the advances, understanding, and humor of these stories. Most notable are my students, fellows, and colleagues who have embraced our ideas enough to help us uncover the many secrets nestled in the genes of wildlife species and in ourselves. Their efforts and determination are the cement that holds these parables together.

Sean Desmond, Thomas Dunne, and Sarah Stewart provided critical editing that consistently improved the reading flow by offering a better way to say it. I am very grateful to my agents, Gabriele Pantucci and Leslie Gardner, for cheerfully guiding me through the perils of trade publishing.

Numerous friends and colleagues freely offered their expert opinions on individual chapters; among these are Karl Ammann, Scott Baker, Margaret Carpenter, Mary Carrington, Victor David, Michael Dean, Theodore Friedman, Murray Gardner, Dianne Janczewski, Warren Johnson, Bailey Kessing, Michael Klag, Lu Zhi, Laurie Marker, Janice Martenson, William Murphy, Craig Packer, Marilyn Raymond, David Reich, Jill Slattery, Mary Smith, David Wildt, and Richard Willing, and Cheryl Winkler. I am grateful to all for their help.

This book would not have emerged without the generous support and enthusiasm of Bob May, Roy Anderson, Jessica Rawson, and Richard Southwood, who hosted my sabbatical retreat at Merton College, Oxford University, where the first chapters were composed. For their cheerful and stimulating sponsorship I remain in debt.

My sister, Carol Reed, my mother, Kathryn O'Brien, and my wife, ~~Diane O'Brien, critically read an early draft of the~~ entire manuscript. Their insightful, pointed suggestions improved the book considerably. I also thank my good friends who critiqued a more polished version, namely, Eric Lander, Richard Leakey, Ernst Mayr, Thomas Lovejoy, Robert Gallo, and Peter Raven.

Finally, I thank my daughters, Meghan and Kirsten, whose humor and curiosity motivated my own, and whose fledgling intellectual development made this book both timely and necessary.

Stephen J. O'Brien
February 4, 2003

Index

Note: **Bold** references are to the Glossary. Gene names are in *italic*; protein names are in roman.